Making Wooden Mould for Refractory Brick Shapes

Pneumatic Ramming Process

By Sheojee Prasad

Illustrated and Designed by Sheojee Prasad

Rev 1.0 dated 08-Jan-2018

Copyright © 2018 by ReeSaa Pvt. Ltd. (**reesaa@indiavivid.com**)

All rights reserved. Without limiting the rights under copyright reserved above, no part of this publication may be reproduced, stored in or introduced into a retrieval system, or transmitted, in any form, or by any means (electronic, mechanical, photocopying, recording, or otherwise) without the prior written permission of the copyright owner

About the Author

Sheojee Prasad is an Author and practitioner with 40 years of experience in Tata Refractories. As a passionate educator and a professional he offers his experience to educate next generation of career professional in the field of refractories, pattern making and mould making.

In this book, Sheojee Prasad makes the task of pattern and mould making in foundry simple.

To begin with, this book explains the concept of pattern and mould making with practical examples and guidelines.

Also by Sheojee Prasad

1. Pattern & Mould Making in Foundry by Amazon.com

https://www.amazon.com/dp/B018AV495C

2. Pattern & Mould Making in Foundry - Vol.2 (Practical Exercises) by Amazon.com

https://amzn.to/2U3i0yq

3. Engineering Drawing – A Practical Approach

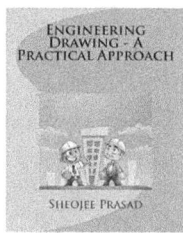

https://amzn.to/2YB3rRh

4. Mould Design Using Hydraulic Press: For Refractory Bricks – The Practical Way of Designing Moulds

https://www.amazon.com/dp/1542573726

5. Jigs, Fixtures and Measuring Instruments

https://www.amazon.com/dp/1797930362

Acknowledgements

I have working experience of about 40 years in multidiscipline functions such as Pattern design, Pattern making, Mould design and moulds manufacturing for production of refractories shapes. I had the opportunity to look after the design and manufacturing of wooden and steel moulds for the various complicated individual as well as assembly shapes.

I offer my thanks to management and my boss for giving me opportunity to work in wooden mould shop, steel mould shop, Machine shop, Flow control mould shop, mould design, manufacturing and refractories lining design

I also offer my thanks to the team in various departments for implementing my innovative ideas for benefits to the company.

I always tried to share my ideas and thoughts with my team and graduate training whoever came in my contact.

I have documented my experience on this book for the benefit of freshers and refractories making units. I am sure it will guide readers.

I'd like to thank Santosh Kumar who has inspired me to document my experience and knowledge on paper. I also thank him for editing this book.

Author's note

This book will take you to some complicated and important expects of wooden mould making. The manufacturing moulds look to be very easy. It is easy of course for simple shape but difficult for complicated shapes. There are certain guidelines and rules that are to be followed.

This book will provide step by step guideline with brief description and sketch. I have written this book to inspire those who are willing to work in refractories shape making organisations and shop floor supervisors of production and maintenance.

Sheojee Prasad

Introduction to this book:

I have written this book as a practical guide for self-learning mould making with wood which is very essential to manufacture complicated refractories shape bricks. This book is full of interesting topics related to wooden mould making to manufacture refractory shapes.

I have described all necessary tools and machines used by mould makers with pictorial views.

I have described step by step the process of making wooden mould and brick making process with pneumatic rammer. It contains the process of making layout on layout board and detail design of all parts of mould.

I have some question and answers to test the skill after going through this book. You are advised not to see the answer directly. First write down your answer then check it.

If you do not have carpentry skill and not conversant with the use of carpentry tools see my book "Carpentry work Guide"

The Topics are easy to understand as associated with Engineering drawings and pictures (Pictorial Views). You can go through my book "Engineering Drawing – A practical Approach" if you want to have clear concept of engineering drawings.

Sheojee Prasad

Contents

Chapter – 1   06
Refractories and area of application

Chapter – 2   10
Mould making steps

Chapter – 3   15
Associated skill, required tools and machines

Chapter – 4   28
Bench work, marking, cutting, drilling, tapping and plate fitting

Chapter – 5   42
Basic knowledge of gas cutting and welding

Chapter – 6   45
Classification and design criteria

Chapter – 7   50
Practical exercises

Chapter – 8   113
Question for skill test

Chapter – 9   114
Answer to questions on chapter – 8

Chapter – 1
Refractories and its Area of Application

Introduction:

Refractories are materials that have very high fusion temperature. It is used in various sizes and shapes to save the structure of metals from deformation and melting down during heating process for various purposes. It is acting as heat resisting barriers between high and low temperature zone. The inner surface of heating equipment where temperature can rise above 500^0C is covered with protecting coating of refractory lining. The refractories can withstand temperature as high as 2000^0 C or even more.

All sorts of heating equipment's shell or structures are lined with refractories material to maintain stability and prolong life while under exposure to intense high temperature during operation. It withstands and maintains internal temperature to required level.

Common area of application

- Steel plants and foundry:
    1. Blast furnace
    2. Converter
    3. Ladle
    4. Tundish
    5. Electric arc furnace
    6. Cupola Furnace
- Glass Tank (Glass Industries)
- Limekiln
- shaft Kiln

- Chemical plants
- Cement plants
- Power plants
- All other heating equipment that is subjected to high temperature during operations.

Refractories Materials

- Fireclay
- Oxides of Alumina
- Silicon(Silica)
- Magnesium
- Dolomite
- Zirconia
- Silicon carbide
- Graphite

Classification of refractories on the basis of stability against slag and prevailing atmosphere in furnace

- **Acidic refractories:** The shapes are not affected by acidic materials
- **Neutral refractories:** These are chemically stable and used in an area where slags and atmosphere are either acidic or basic.
- **Basic refractories:** These are used where slags and atmosphere are basic. They are stable to Alkaline but could react with acids.

Fusion temperature of refractories

Grade of refractories	Fusion temperature
Normal refractories	$1580 - 1780^0$ C
High refractories	$1780 - 2000^0$ C
Super refractories	More than $- 1780^0$ C

General products of Refractories Industry

- **Brick shapes**
    1. Standard size bricks
    2. Nonstandard-size bricks
    3. Shape bricks such as for coke oven
    4. Special shape bricks such as bottom pouring refractories
- **Monolithic**
- **Ramming mass**
- **Gunning mass**
- **Mortar**

Product range on the basis of raw material and composition

- Silica
- Basic
- Fireclay
- High alumina
- Dolomite
- Special refractories.

Formation of refractories shapes

The manufacturing of brick shape needs special effort through various processes such as

- Process design / planning
- Selection of raw material
- Setting the composition to suit customer's need

Process design and planning is done to take decision on

- Mould making
- Setting the composition

- Mode of formation of the shapes to suit the application need at customer's end such as machine moulding, hand moulding, precast shape etc.
- The bulk density of shape considering the application.
- Selection of press
- Dry chamber processing.
- Processes of firing
- Inspection

Chapter – 2

Mould Making Steps

Mould:

It is an object having inner and outer shapes. The inner shape is formed by mould maker with necessary allowances according to brick shapes and sizes. In case of simple shape, it will be fully or partially negative shape of the brick to be made. In complex shape of bricks having features on top and bottom surface it is associated with top and bottom dies. The outer and inner shapes are governed by various factors that will be clear in due course. The moulds are used for making bricks with suitable refractories mixture for specific application in heating equipment.

Process design for making bricks:

The process design is the responsibility of Research and Development Laboratory to set the composition to suit the customer's requirement. The person concern will take decision on

- Suitable composition
- Expansion or contraction allowances
- Setting, drying and firing
- Mode of forming the shape

Planning:

The process design and planning is very important to manufacture refractories bricks to suit the customer's requirement. The concerned person will consider

- Delivery period
- Manufacturing processes comprising of
    1. Selection of mode of pressing such as Press, Precast or Ramming
    2. Mould design and mould making time
    3. Manufacturing time
    4. Setting, drying and firing time
    5. Inspection

The planning is very important and knowledge oriented work. The planner must have knowledge of manufacturing process and <u>clear concept of engineering drawing.</u>

Mould Making:

The first step in mould making is to design the mould on the basis of the following information:

- Bulk Density of shape
- No of pieces to be moulded
- Drawing of shape to consider ease of ejection or removal from the mould.
- Quality of brick with expansion allowance or shrinkage allowance that is considered on mould dimensions.
- Assembly or non-assembly shapes
- Mode of inspection by customer
- Pressing mechanism.
- Press capacity and Specific pressure to be applied for taking decision on no of cavity in case of steel mould for hydraulic press.

Category of moulds on the basis of mould material, the intricacy of shape and no of pieces required:

- Wooden mould.
- Steel mould

Wooden Moulds:

The major part of the mould is made of wood. The surfaces that come in contact with brick mixture are covered with B.P. sheet. The seasoned teak wood is selected for making wooden moulds. The factors involved in taking decision on making wooden moulds are

- **Complicacy of shape:** The shapes that can't be ejected from the press moulds are planned for detachable wooden moulds even for higher production.
- **No of bricks to be made:** Generally, small quantities are considered for wooden moulds.
- **Bigger sizes:** The brick size that could not be made on hydraulic or mechanical press.
- **For making precast shapes:** The shapes that are not possible to be made by hand moulding or pressing on the press. The decision is also taken on the basis of quality requirement. Examples burner blocks, glass hearth blocks, Tuyere, Bustle pipe bends, hot blast Main-Bustle pipe junction in Blast Furnace, Carbon Block etc.

Detachable mould:

The detachable mould is the type of mould assembly that can be easily assembled for moulding and dismantled after ramming brick mixture to take out brick from mould.

Steel moulds:

The various grades of steel with hardness are used for making steel moulds. The factors involved in selection of steel, decision on heat treatment and hardness depends on

- No of bricks required to produce
- B.D. of bricks to suit the application
- Quality of bricks and abrasiveness of composition.

The shapes that could not be ejected are not planned for the press.

Mould design:

The mould design is very important in the formation of refractories shape. The critical shapes also can be made in the press with a suitable design of the mould. <u>The mould designer must have a clear concept of engineering drawing. The minor mistakes in mould design may cost heavy on product and finally on the company.</u>

Wooden Mould Shop:

The mould shop for making wooden moulds is one of the very important wings of manufacturing refractories shapes. It deals with various complicated individual shapes as well as assembly shapes that couldn't be made with a press.

Works involved in mould making activity:

- Procurement of suitable wood in association with material management and inspection department. This work is done by mould shop where there is no material management and inspection department. The inspection of quality is being done by mould shop in presence of inspection authority.

- Procurement of steel plates, nails, screws, ties bolts, nuts etc. through material management.

Wood and associated materials for making wooden moulds

- Wood: The Teak wood, Mahogany, Haldu wood, Pinewood etc. in various length, width and thickness are procured for use as required by individual refractories plant.
- Screws: countersink headed screws of various sizes.
- Hexagonal bolts and nuts of various sizes
- Allen bolts and nuts of required sizes

Suitable wood for Making moulds:

The seasoned Teak Wood is regarded as best suitable wood for mould making. It has minimum bending tendency as such deformation in size and shape is considerably less than other woods.

Screws: The screw is the type of fastener made of metal. It has external helical thread. It has slight taper from body to tip. It has pointed end. The helical thread starts from pointed tip and ends at straight portion of the body.

Chapter – 3

Associated Skills, Required Tools and Machines

Associated knowledge and skills for wooden mould making:

- Simple mathematical calculations
- Knowledge of engineering drawing
- Carpentry skill
- Bench work (plate fitting)
- Welding
- Gas cutting
- Soldering.
- Brazing

Primary requirements of a wooden mould maker

- Working bench
- Power-driven and manual tools
- Machines

Working bench:

The working bench is very essential aid of carpentry works and mould making. It should have comfortable height and tools storage space. The wooden planks are placed on the upper surface of the table and set against stopper made of wood having steel pin to hold the workpiece. The carpenter's vice is also associated with it to hold the workpiece for required operations. See the drawing of working bench with carpenter's vice.

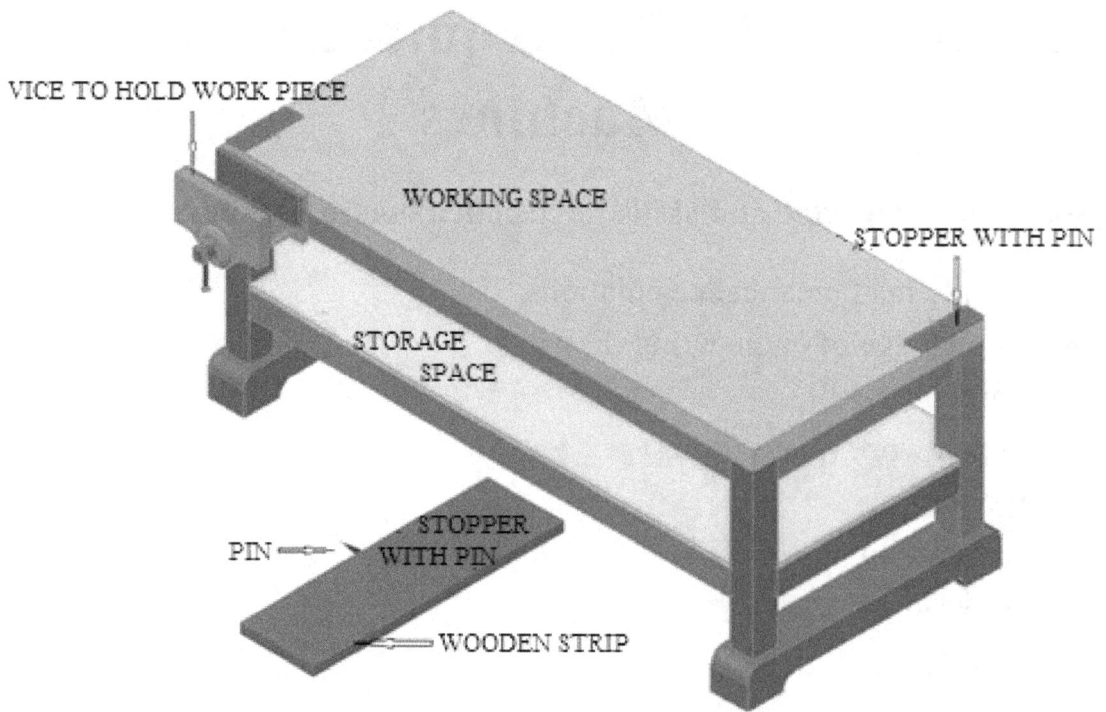

The list of hand tools:

- Power-driven portable planer
- Hand Planer
- Wood cutting tools
- "T" clamp : It is T shape stoper with long guide bar with round holes at reguar intervals to move and lock movable stopper. It is used for components assembly of larger size than "C" clamp.
- "C" Clamp: It is used for clamping work piece on working bench and for smaller assembly

- Spanners and pliers
- Drill bits
- Callipers (inside and outside)
- Bevel protectors
- Auger Bits
- Compass
- Beam compass or Trammel
- Gauge or Round chisel for cutting round profile
- Screwdriver (power driven or manual)
- Spanners
- Ratchet
- Hammer
- Mallet
- Handsaw
- Files
- Marking Gauge.

Manual operated hand tools:

The pneumatic ramming moulds are made with wood. All the wood working tools used by carpenters are also used by mould makers. Read my book "carpentry skill guide" to learn the carpentry skill.

Machines required by wooden mould shop:

- Wood planning machines
    1. Surface planner
    2. Thickness Planer
- Circular Saw
- Plate shearing machine
- Plate bending facilities (Bending machine or dies for manual bending)
- Wood cutting bandsaw

- Bench grinder
- Drill machine
- Disc Sander

Portable Circular saw:

The circular saw and its baldes are designed to suit the cutting requirement of wood. The drawing drawn here is just to illustrate the fundamental working principle of circular saw. The operating system of circular saw may vary according to design and need to cut the stock.

The circular saw blades are designed to suit the material to cut. It is designed to cut along the grain of wood (called rip-cuts), to cut across the grain of wood(called cross-cuts) or a combination of both

The blde of the saw will have rotary motion in clockwise direction when it is pluged with electic power. The base plate will touch the work piece that is to be cut. The guard is adustable to have suitable leverage on stock. The depth of cut also adjusted by moving up and down blade with respect to base plate. The blade can be replaced with suitable design of cutting teeth profile.

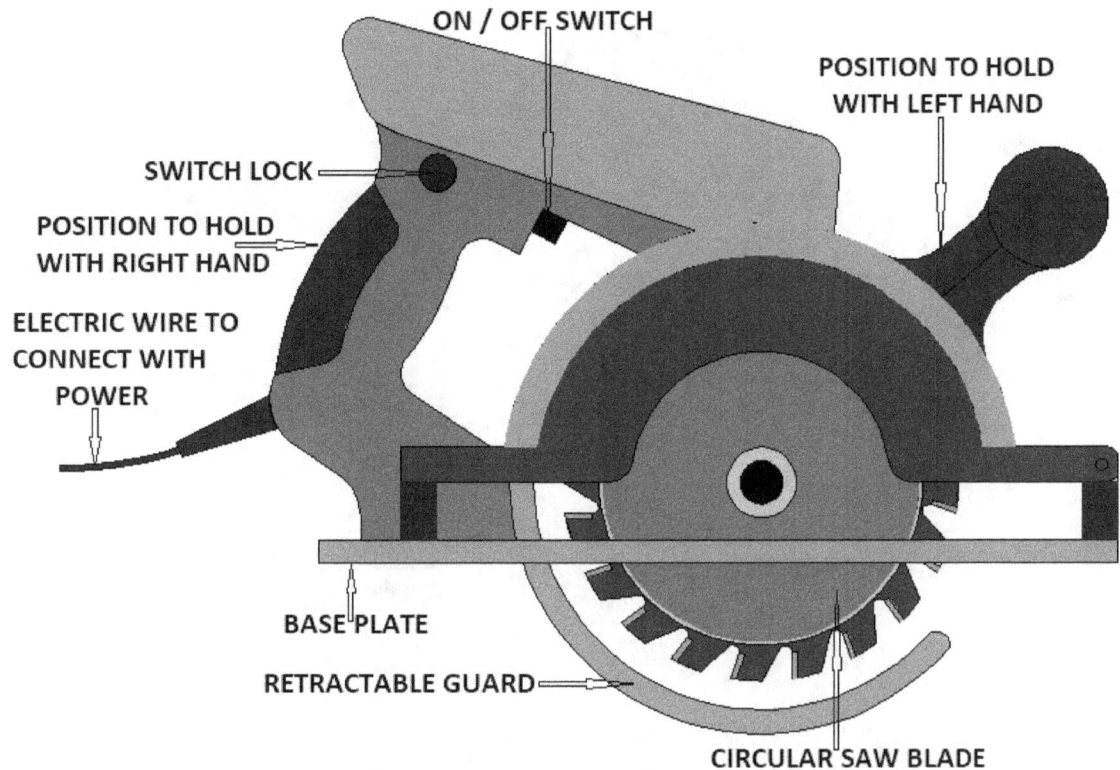

Sketch of bandsaw machines and part details:

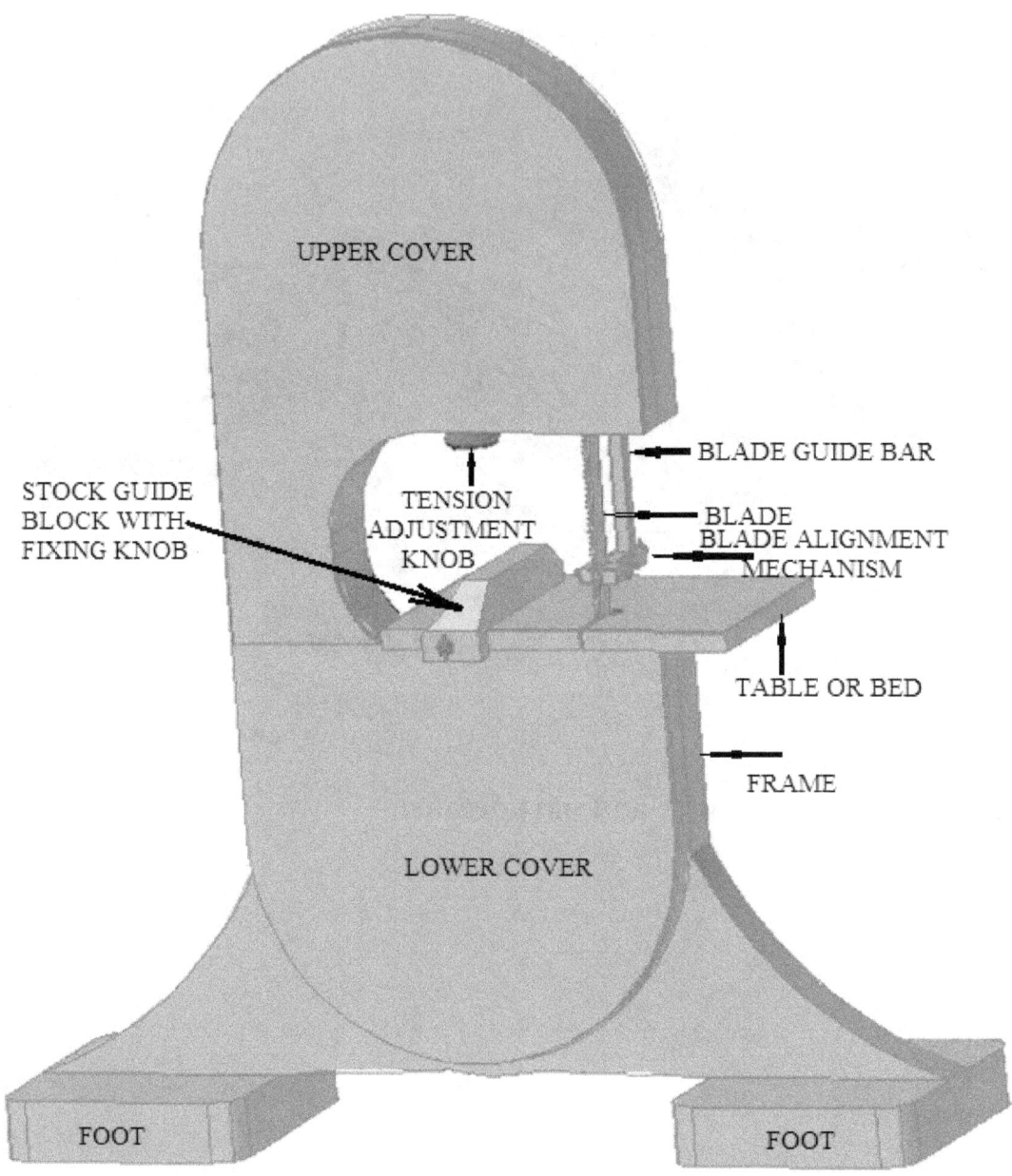

Note: Observe carefully there is a circular block just near the back end of blade on the table; It is throat plate of wood or hard rubber to guide the blade. The cover plates have been removed in the figure given below. It has hinge attachment with frame.

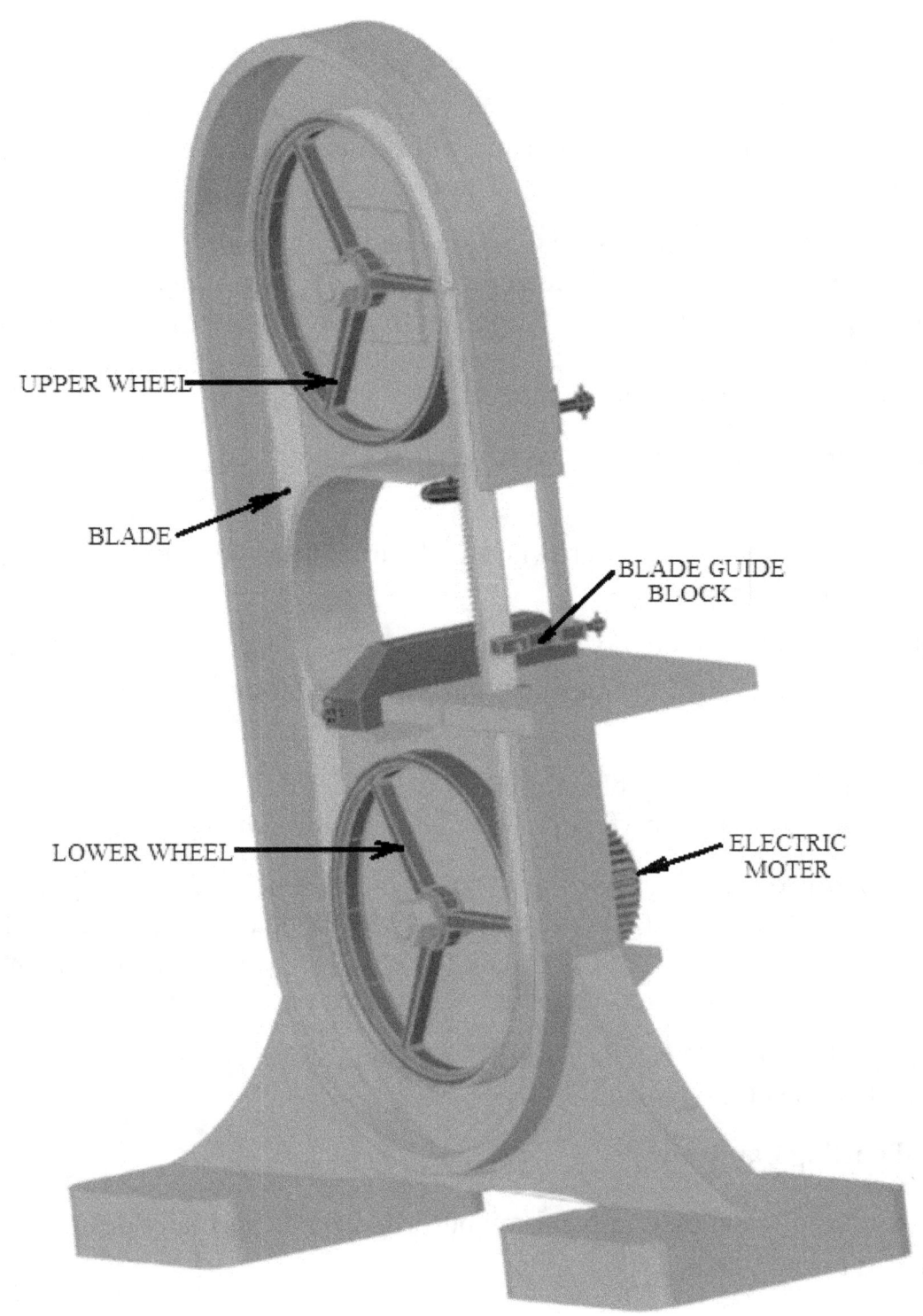

Parts of wood cutting Bandsaw:

Frame: The frame is the structure made of cast iron to support components of bandsaw such as wheels, tension adjustment mechanism, blade guide mechanism etc.

Upper wheel: The upper wheel is attached with shaft and bearing in a housing to rotate along with lower wheel. It is also attached with blade tensioning mechanism. When knob is rotated clockwise it aligns and provides tension on blade. It is rotated anticlockwise to loosen the blade while taking out for sharpening the teeth.

Lower wheel: The lower wheel is connected to motor with bearing and shaft to drive it.

Tires: The both wheels are crowned with rubber tires that provide traction for the blade to run on.

Blade: The blade is made of high carbon steel with hardness of RC60 – 61. The teeth angle is same as ripsaw teeth. The setting is also same as ripsaw teeth.

Blade Guide Block: The blade guide block has special mechanism to keep blade always aligned in perfect position while cutting the stock (wood).

Table: The table has all sides and surfaces machine finished to place the stock in horizontal position while cutting. It has elongated slot to insert blade and circular hole to insert throat plate made of wood to guide and position the blade. It also has an arrangement to swivel at required angle.

Stock guide block: It can be fixed at required position on table. It supports the stock while cutting

Metal cutting bandsaw:

It is similar to wood cutting bandsaw but it has blade specially designed and hardened to cut metals.

Bandsaw blade grinder:

The triangular file can be used for Bandsaw Blade grinding (sharpening the teeth) in non-availability of bandsaw blade grinder.

Bench Grinder or pedestal grinder:

It is used for grinding cutting tools

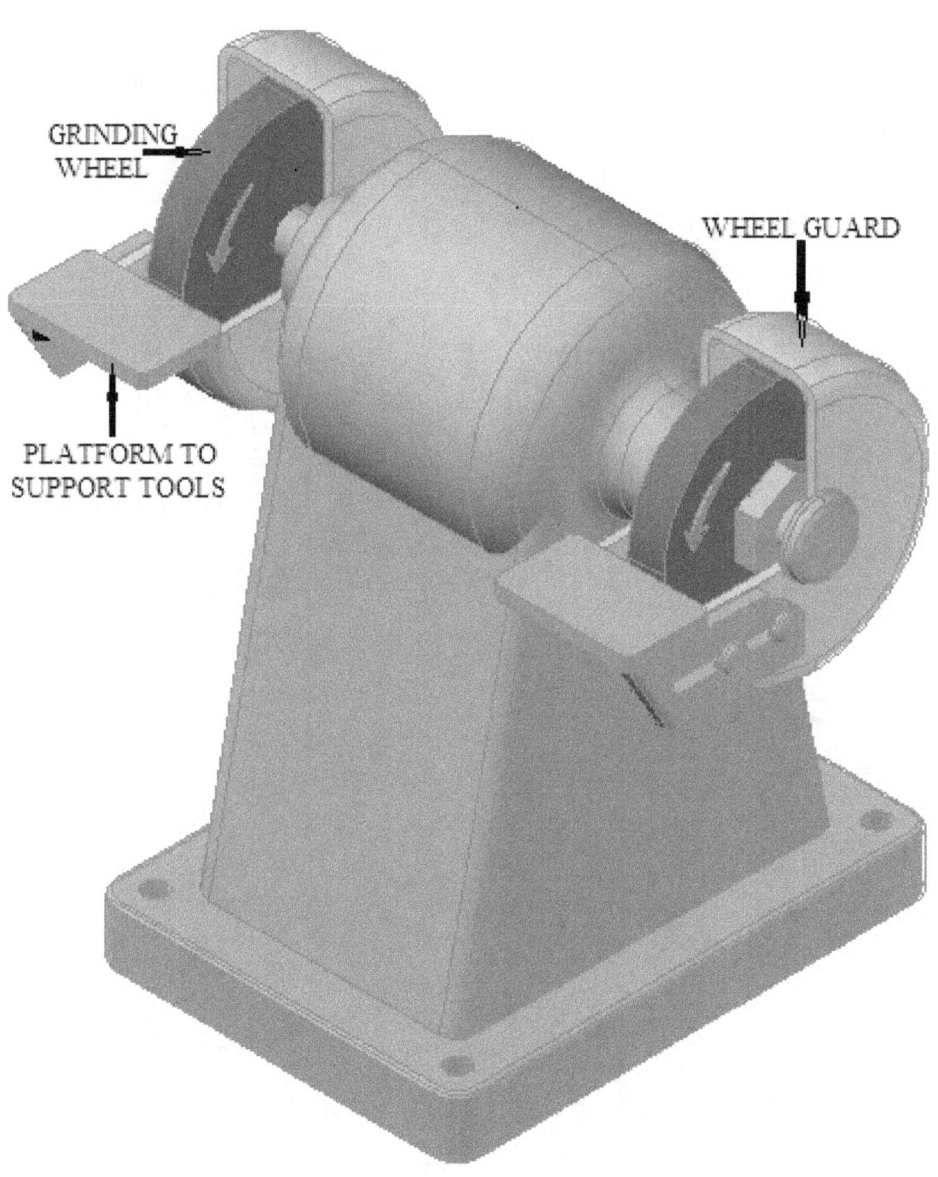

Pillar Drill Machine

The drill machines are used for drilling required holes on wood and metal plates.

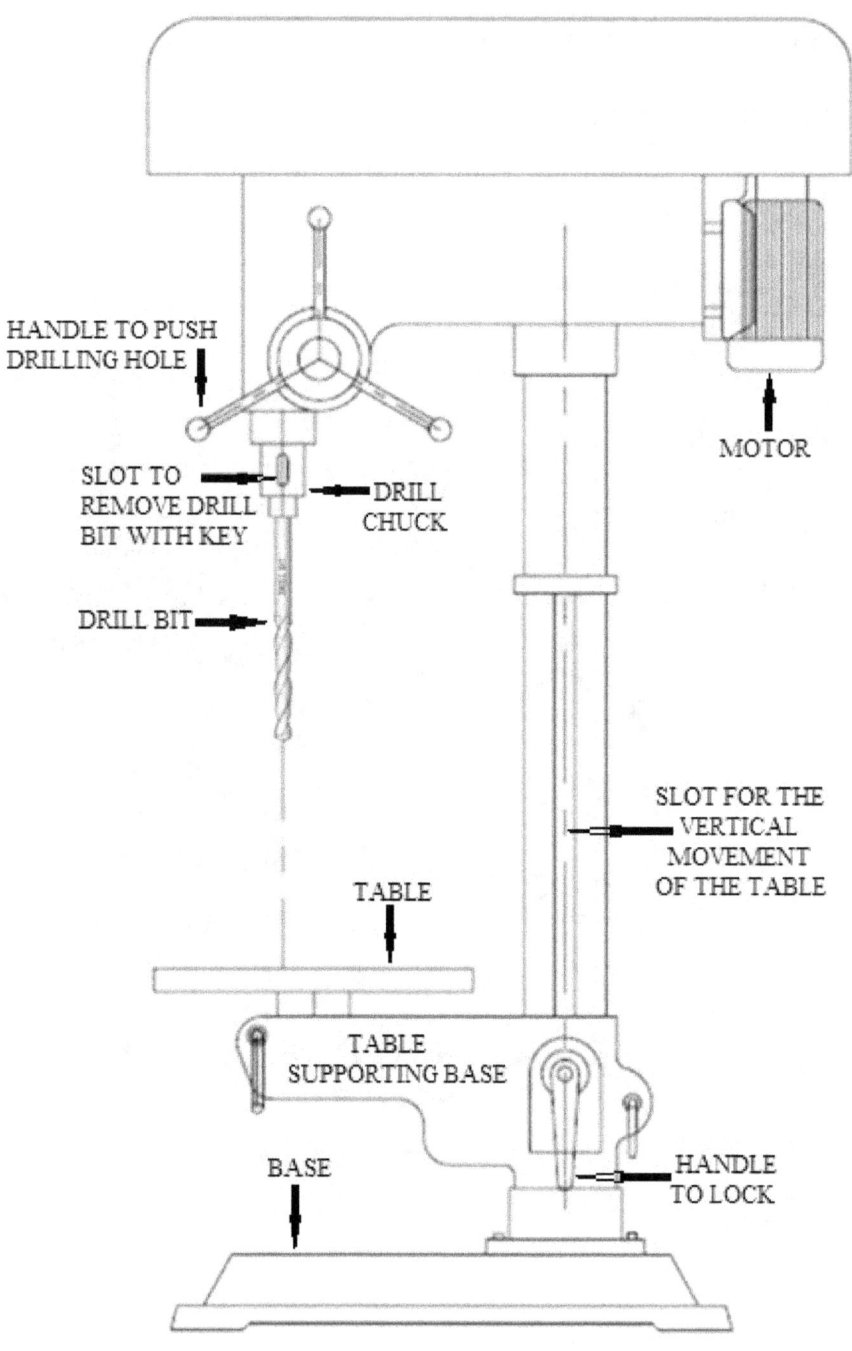

Important Power driven portable Tools:

Portable hand drill machine:

The portable drill Machine is very essential tool for carpentry work. The capenter have to make holes of various sizes at different locations. The drill is tightened with drill chuck after inserting in the jaw. The sketch is self explanatory.

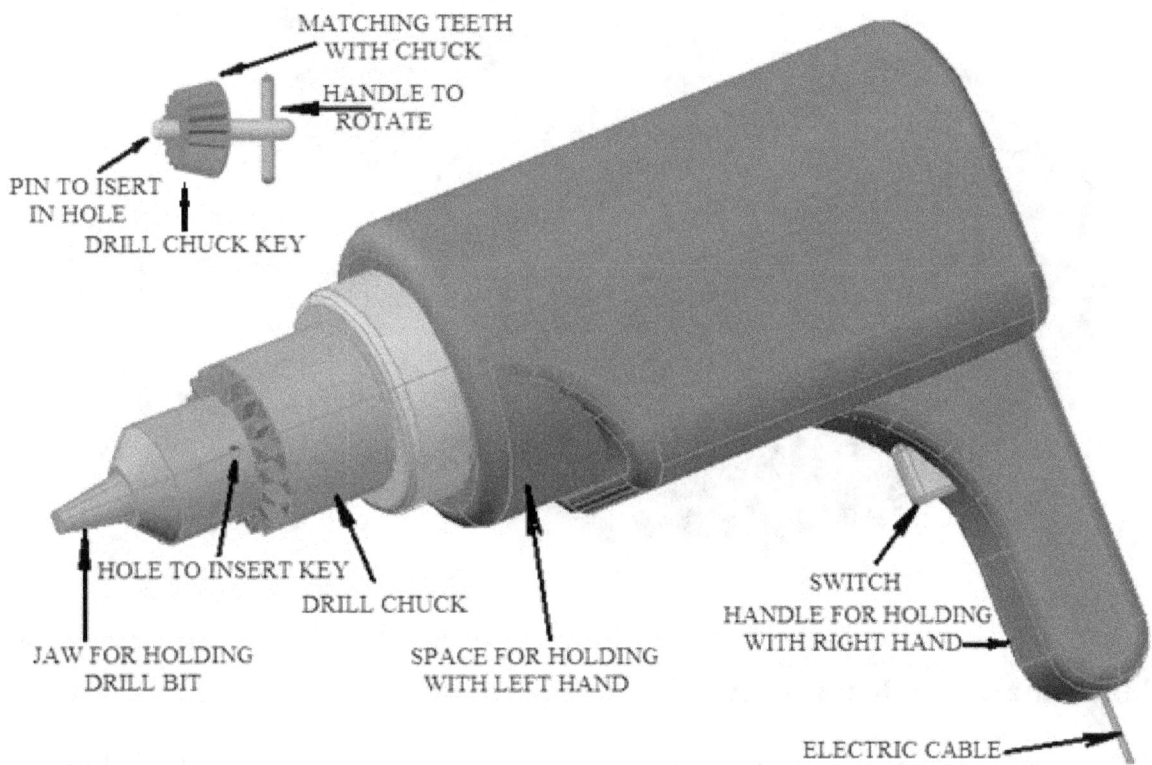

Drill bits:

Different sizes of drill bits are used for making holes in wood or other materials. The holes are made with the drill bits in association with drill machines or portable hand drill.

Disc grinder:

The disc grinder is used to grind rough surface to make it level and smooth. It works faster than other hand tools.

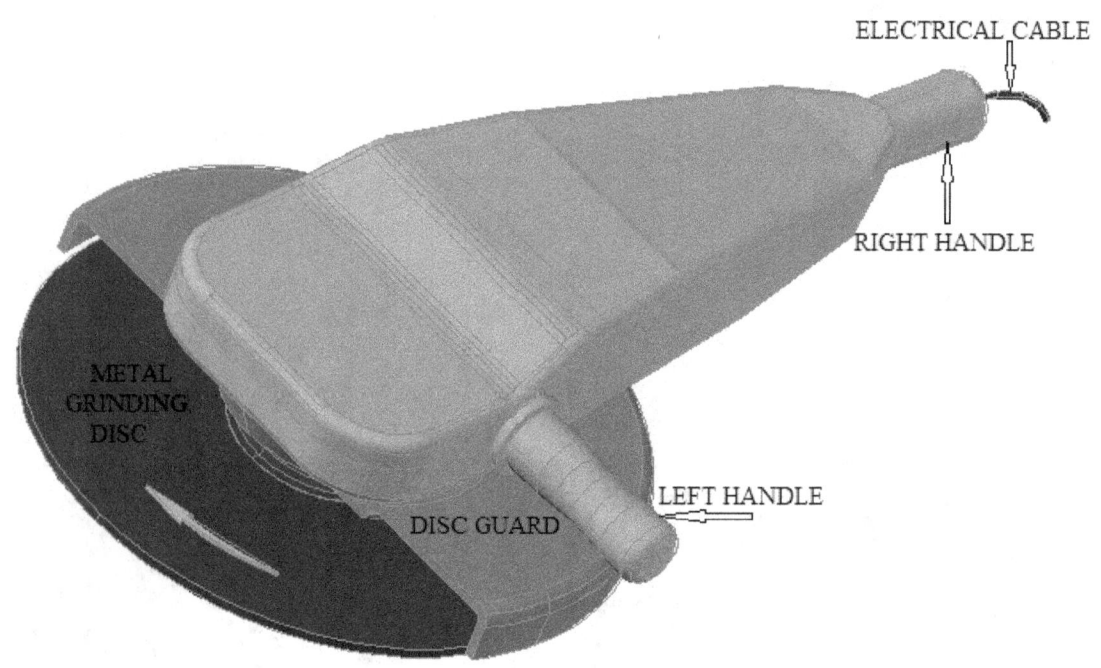

Mathematical calculations:

The mould maker must have the knowledge of conversion table, addition, subtraction and multiplication of whole numbers, fraction and decimal numbers. He should also know percentage calculation for making moulds with shrinkage and expansion allowances.

Examples mould allowance calculations:

Expansion Allowance:

Find the size of mould to be made with 4.5% expansion allowance for making brick of size 600 * 450 * 325 mm

4.5% expansion means the green brick with 100mm measurement will expand to 104.5mm after firing and Cooling down to room temperature. The green brick means bricks taken out from mould cavity after compressing the mixture.

Under above condition for 104.5 mm brick size in drawing mould size will be 100mm, mould size for 600 mm will be (600 * 100)/104.5 = 574.16 mm

In reverse way 574.16 will become (574.16 * 104.5)/100 = 599.9972 = 600mm

Similarly mould size for 450mm = (450 * 100)104.5 = 430.62mm

And for 325mm = (325*100)/104.5 = 311.004 mm

The correct mould size for 600 * 450 * 325 will be 574.16 * 430.62 * 311.004

Contraction (Shrinkage) Allowance:

Find the mould size for brick of 650 * 400 * 300 with 5% Shrinkage Allowance

The Shrinkage allowance means brick will have less measurement on linear size by 5 mm on every 100 mm after firing. It means you should make mould with 100 mm when drawing size is 95 mm. The brick with 100 mm measurement shrinks to 95 mm after firing and cooling down to room temperature.

Mould size for 650 with 5% Shrinkage will be (650 * 100)/95 = 684.21mm

Mould size for 400mm with 5% Shrinkage will be (400 * 100)/95 = 421.05 mm

Mould size for 300mm with 5% Shrinkage will be (300 * 100)/95 = 315.789 mm

Check the calculation in reverse way when 100 will shrink to 95mm then 684.21 will shrink to (684.21 * 95)/100 = 649.9995mm.

Knowledge of engineering drawing:

The knowledge of engineering drawing is very essential for making shape bricks. You may go through my book "Engineering Drawing (A practical Approach)" if you lack knowledge. There will be no any problem to make moulds for standard brick sizes with linear measurements but to make moulds for critical shape bricks drawing knowledge is required.

Chapter – 4

Bench work (Plate fitting):

Introduction:

The bench work for mould making involves filling, drilling, tapping, plate cutting, and plate bending. The person concern must be fully conversant with use of various types of files, drill, drill size for corresponding tap size, dies for plate bending, sizes of bolts and nuts, specification of nut and bolts

Machines used by Plate fitting group:

- Shearing machine
- Plate bending machine
- Metal cutting bandsaw
- Drill machine
- Tool grinder (pedestal grinder)
- Disc grinder
- Portable drill machine

Hacksaw: The hacksaw is used for cutting plates with hand

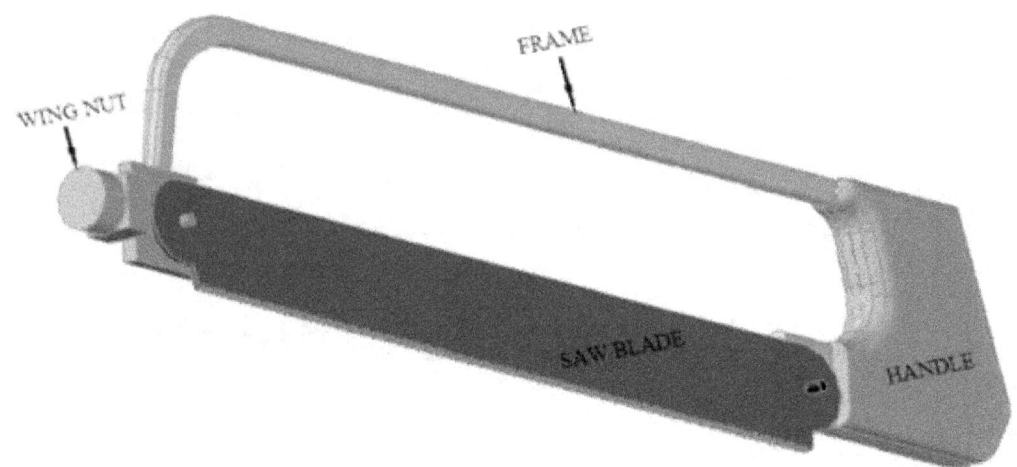

Cold chisel and hammer:

The cold chisel is used to cut sheet with a hammer. It can cut any shape and size of the plate.

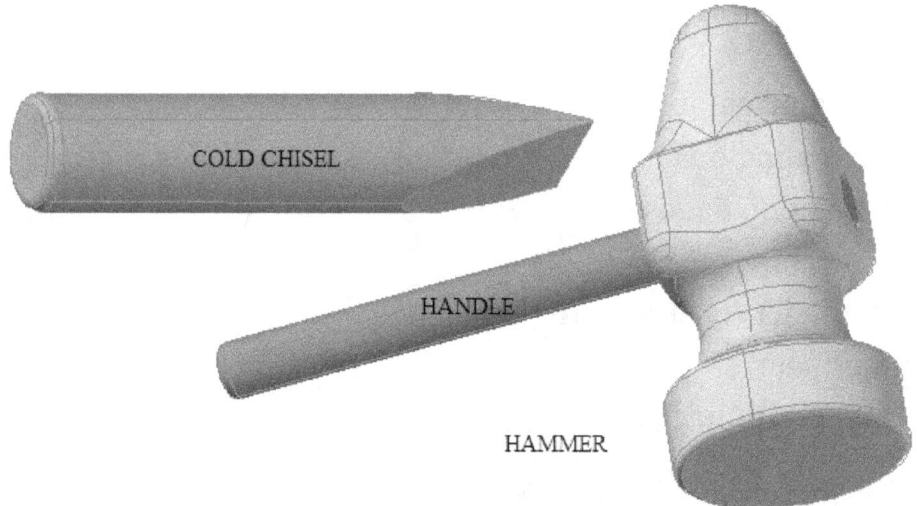

Punch:

The punch is used to mark dots over lines on the surface of plate or point for drilling holes. It is done for prominent visibility during working with sheet metal.

Countersink Cutter:

D is the diameter of the hole to have countersink. The tip angle of the cutter is 90^0 that will fit with a wood screw.

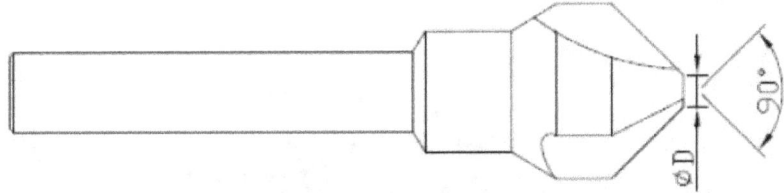

Internal thread cutting Taps:

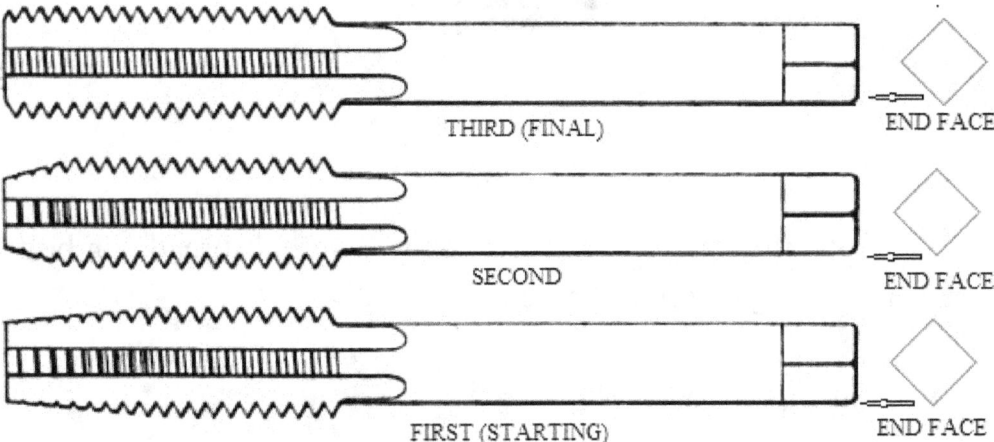

Taps and Tap wrench:

The taps in association with tap wrench are used for making internal threads in holes according to requirements.

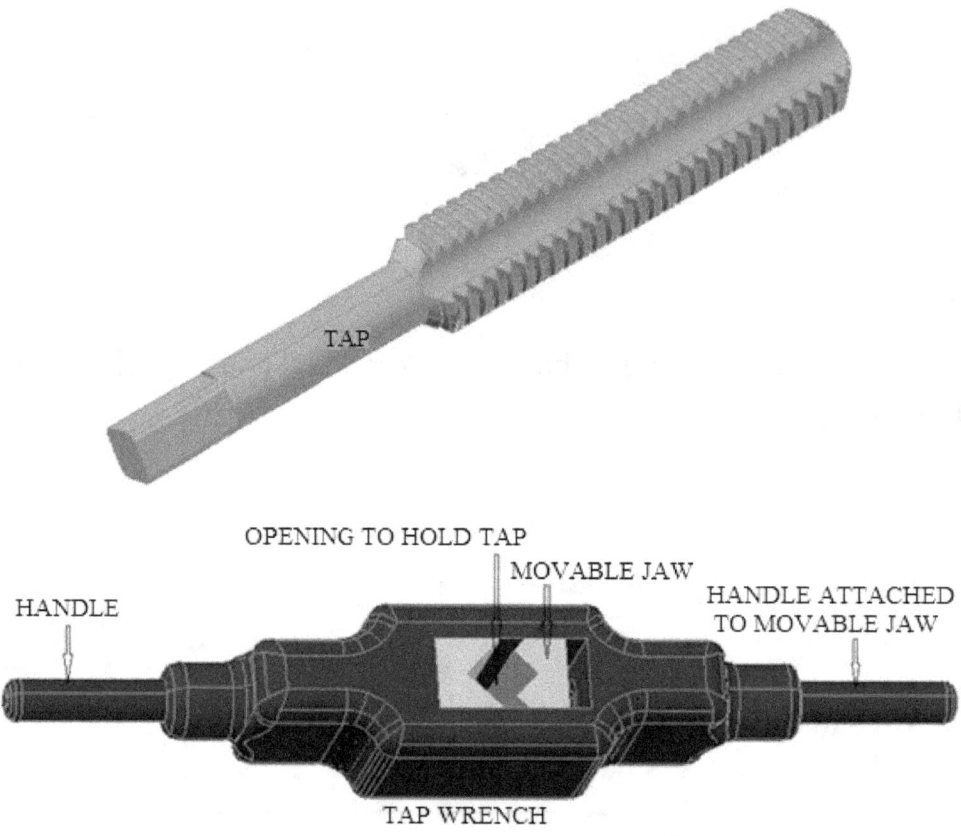

Dies:

The various sizes of dies are used for making external threads on bolts. It is also used for cleaning external threads.

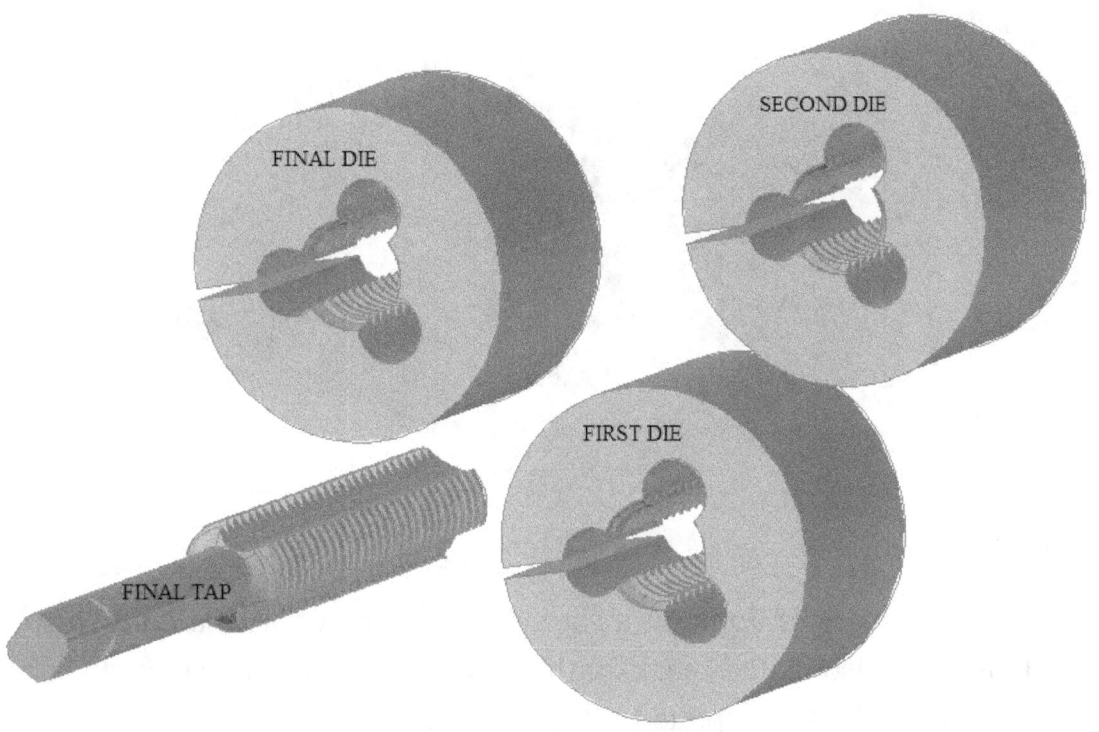

Marking on Plate:

White chalk or blue paint is applied on the surface of the plate. The measurements are transferred from layout and lines are marked with a scriber. The divider is used for making any curve line. Mark a centre with punch at required location and place one leg of divider after taking the measurement of radius from layout out or scale and mark the circle. Any shape or size you can draw with the right angle, divider, set square or marking block on the surface of the workpiece.

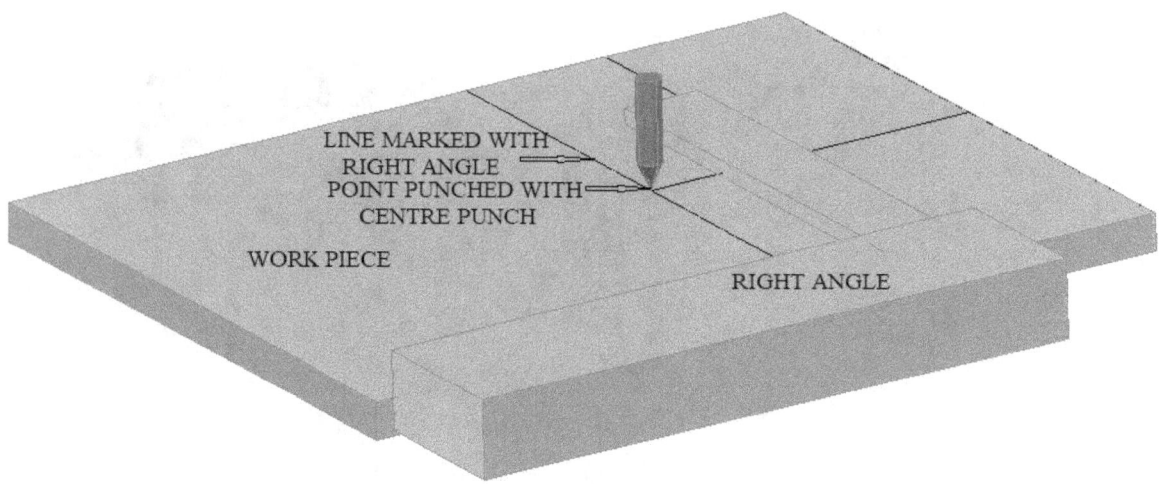

Plate cutting:

The plate cutting work is done on a shearing machine or with a cold chisel. The cold chisel is also called blacksmith chisel. The required size and shape of the plate are marked on a plate by transferring measurements from layout drawing. In manual cutting with a chisel, the cut mark is formed just by the side of lines and bending force is applied. The plate shears off in two pieces with bending force. In case of any special shape cut, it is performed with chisel, hammer or hacksaw. The thin plates are cut with chisel but thicker plates are cut on bandsaw or hacksaw machine.

Cutting with Hacksaw:

Hold the work piece in the vice and cut with hacksaw holding with left and right hands. The left hand will be in front and right hand in the back. Apply forward push force with right hand and press the hacksaw against work piece with left hand. The left hand will also balance the push force applied by right hand. You have to apply pull force also on hacksaw after cutting metal during push force. You left hand will balance the pull force keeping blade in slot formed during push stroke. The left and right hand must move together in synchronise order.

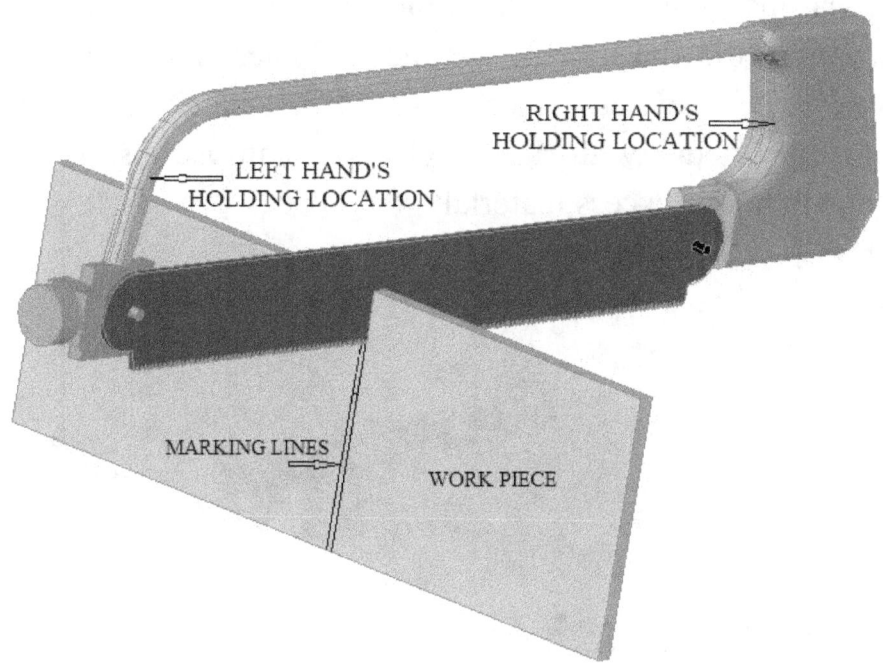

Cold chisel:

It is made of high carbon steel. One end has knife edge and opposite end is flat to receive the impact force of a hammer. The body may be round, square or hexagonal shape; the knife edge tip must be hardened and tempered to make it capable to cut the plate. Hardening and tempering are done by a blacksmith. The chisel is heated and quenched in cold water for hardening and tempering.

Cutting with Cold Chisel:

Wear hand glove in left hand and hold the cold chisel vertically on marked line. Hold hammer in right hand and tap on the head of cold chisel till prominent scar marks appear. Reverse the plate on "v" Channel and hit with hammer keeping cold chisel on line in the back face. The plate will shear in two pieces. File the cut face with file, check straightness and perpendicularity of edge with right angle. You can use disc grinder also if plate is long to remove excess material.

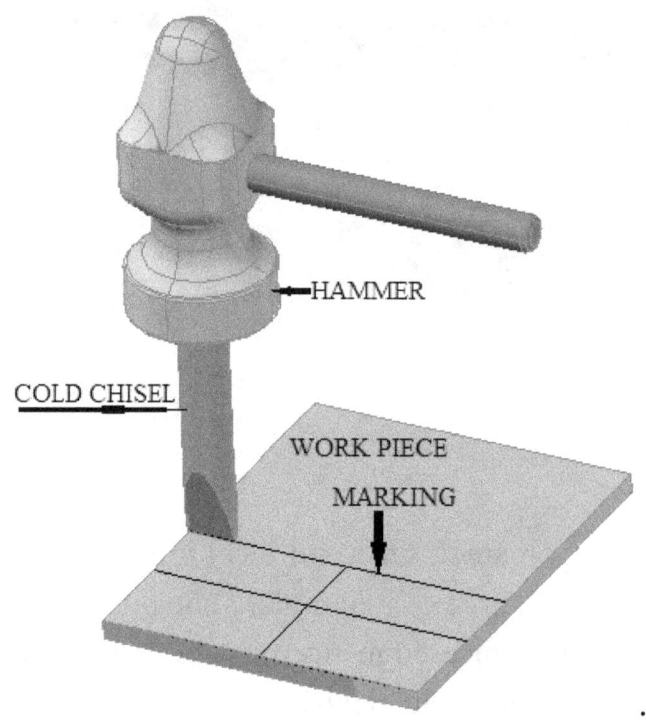

Plate bending tools and fixtures

Mallet and Hammer:

The mallet is used to hit and level the surface of plate gently. It is used for assembly where light push force is required. Hammer is used for cutting and bending sheet meal. It is used where heavy impact force is needed.

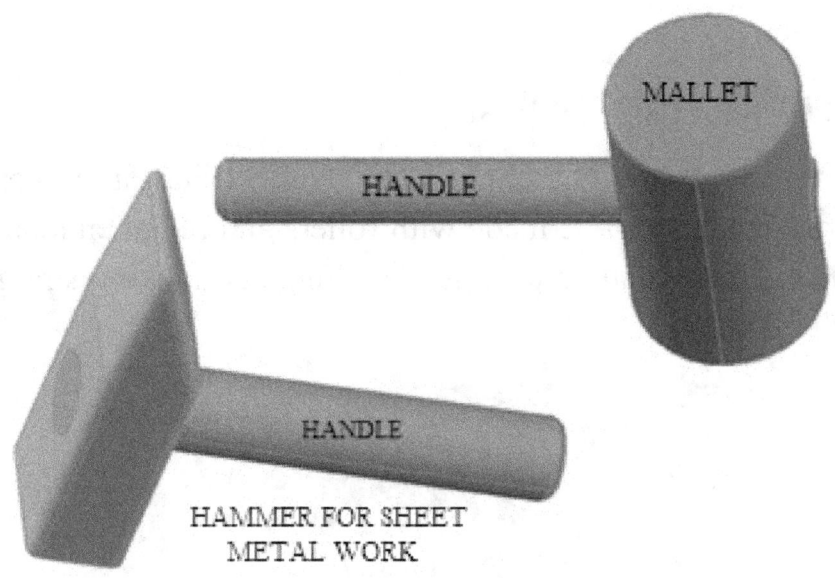

Roller:

It is used for making circular bends of various dimeters. Hold the plate on the surface of roller and hit / tamp with mallet or hammer slowly to get the desired bend. Make template (gauge) with thin sheet for checking the diameter and check during bending process and after bending.

V" Block for right angle bending:

It is used for bending right angle plate with radius at corner. Various required size of radius can be made with rollers and radial end hammer. The back and side surfaces can be used for levelling and making straight surface of plate with mallet.

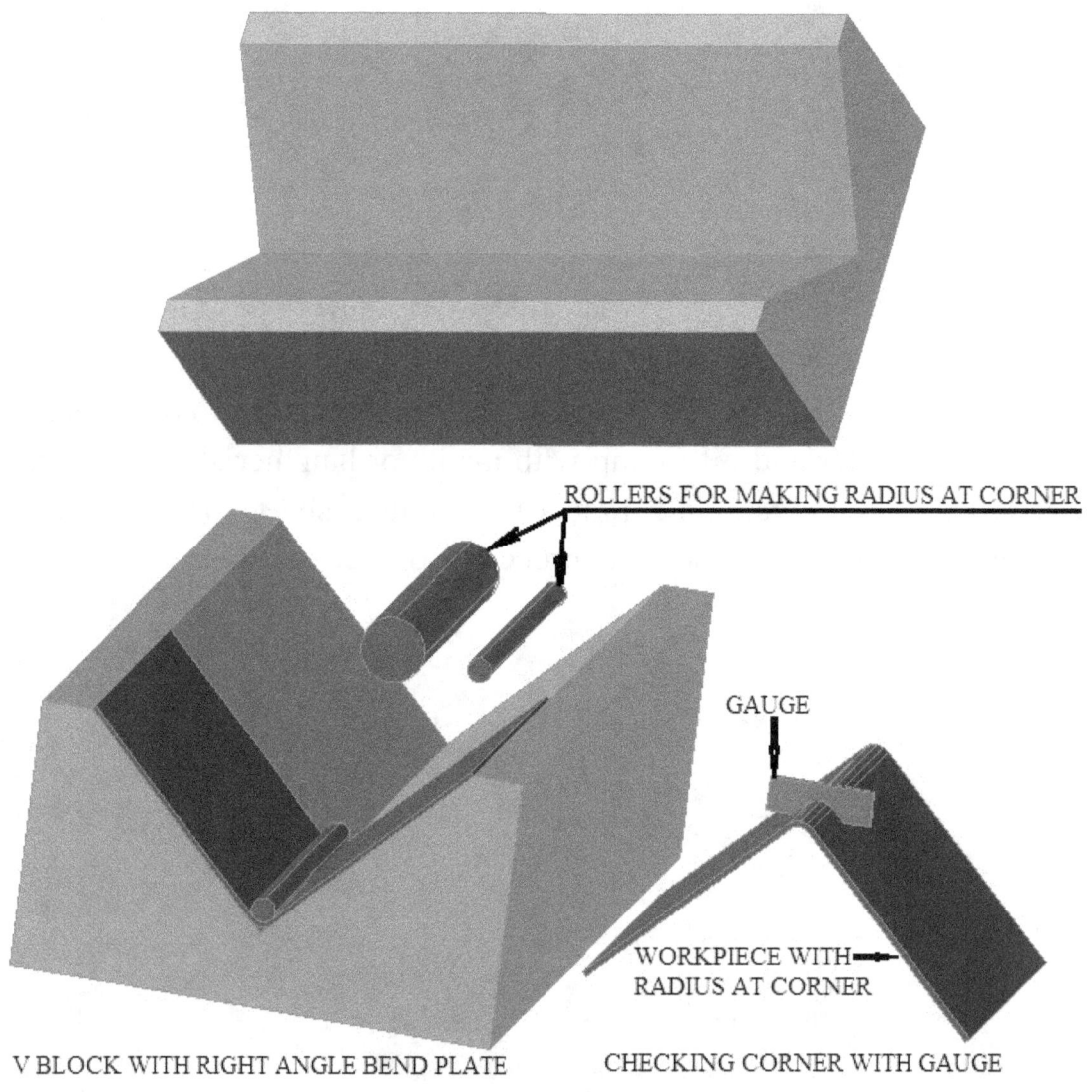

ROLLERS FOR MAKING RADIUS AT CORNER

GAUGE

WORKPIECE WITH RADIUS AT CORNER

V BLOCK WITH RIGHT ANGLE BEND PLATE   CHECKING CORNER WITH GAUGE

Making size and shape:

The suitable size, shape and grade of files are used for filing to make the required size and shape of plates to match with the profile of wooden pieces. The edges, corners and special shape are made right angle with respect to the surfaces.

Fixing plates with wood:

The plates are fitted on the surface of wood that will come in contact with the refractory mixture while ramming to get the brick shape. The countersink holes are made on plates with a drill to insert screws. The suitable countersink screws of required sizes are used to attach with wood through countersink holes. The surface of screw head must flush with the surface of plate.

Drilling and tapping:

The requirement of drilling holes is very common in wood working. The hand operated power driven portable drill is used for making holes in wooden pieces as and when required to join with screws. Making countersink hole on plates is done on drill machine. The head of screw must flush or match with the surface of the plate. The screw selected for fitting is taken as a gauge to check each and every hole while making a counter drill in the plates.

Drilling and tapping holes:

Sometimes it is required to drill and tap holes to fit a specific size of bolt. The person concern must have the knowledge of drill size for required size of tap hole to fit with bolt.

There are three kinds of ISO metric thread

- Coarse
- Fine
- Extra fine

The thread angle is 60^0 and thread depth is 0.614 * Pitch. The metric threads are designated with the capital letter M with an indication of their nominal outer diameter associated with a pitch.

Example: M10 * 1, 5 where 10 is the nominal outer diameter and 1.5 is the pitch of the thread. You can get the drill size to have required thread on as mentioned here.

Tap drill size = major diameter of Tap (outer nominal diameter) – pitch of the tread

Suppose you want to make a metric thread of M10 having 1.5 pitch, then your drill size will be equal to 10.0 – 1.5 = 8.5 diameter.

In case of inch size thread, you can compute the drill size

TD = MD – 1/ N, where TD is tap drill size, MD is the major diameter of the tap and N is the thread pitch.

Drilling holes:

The straight holes are made with drill bits made of high-speed steel. The tip of drill bit point angle is 118^0. The flat ended drill bit with no point angle is used for drilling flat bottom hole.

Tips for drilling holes:

The drill hole must be perpendicular to the surface of the workpiece. When drill machine is used with holding device on the bed of drill machine, the drilled holes will be perpendicular to the surface of the workpiece. But holes made with portable drill machines may be inclined to the surface of the workpiece. Take the help of right angle to eliminate such possibility. Push the drill slowly in the workpiece and check the perpendicularity of the drill bit with the right angle. Place the stock of right angle on workpiece keeping blade vertically up and touch the body of the drill bit. If the drill is going vertically down there will not be any gap in between bit and blade. The suitable size of right angle to be used

Thread cutting in drilled holes:

The threading taps are available in the set of 3 pieces. The first tap (starting Tap) is also called taper tap. It has a higher taper at starting end. This feature enables to make vertical threads at a right angle to the surface. The tap is slowly pushed inside the drilled hole with one rotation forward and half rotation backwards. The taps are used for making internal threads in drilled hole. See the figure that illustrates the tapping of blind hole in the workpiece.

Specifications of counter sink socket screw, washer, Hexagonal Bolt, and nut:

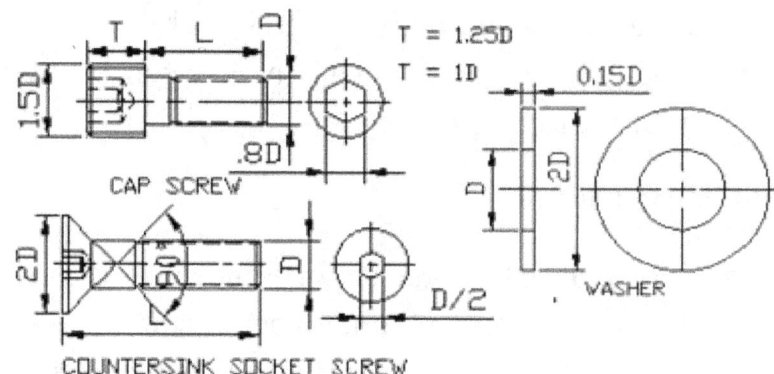

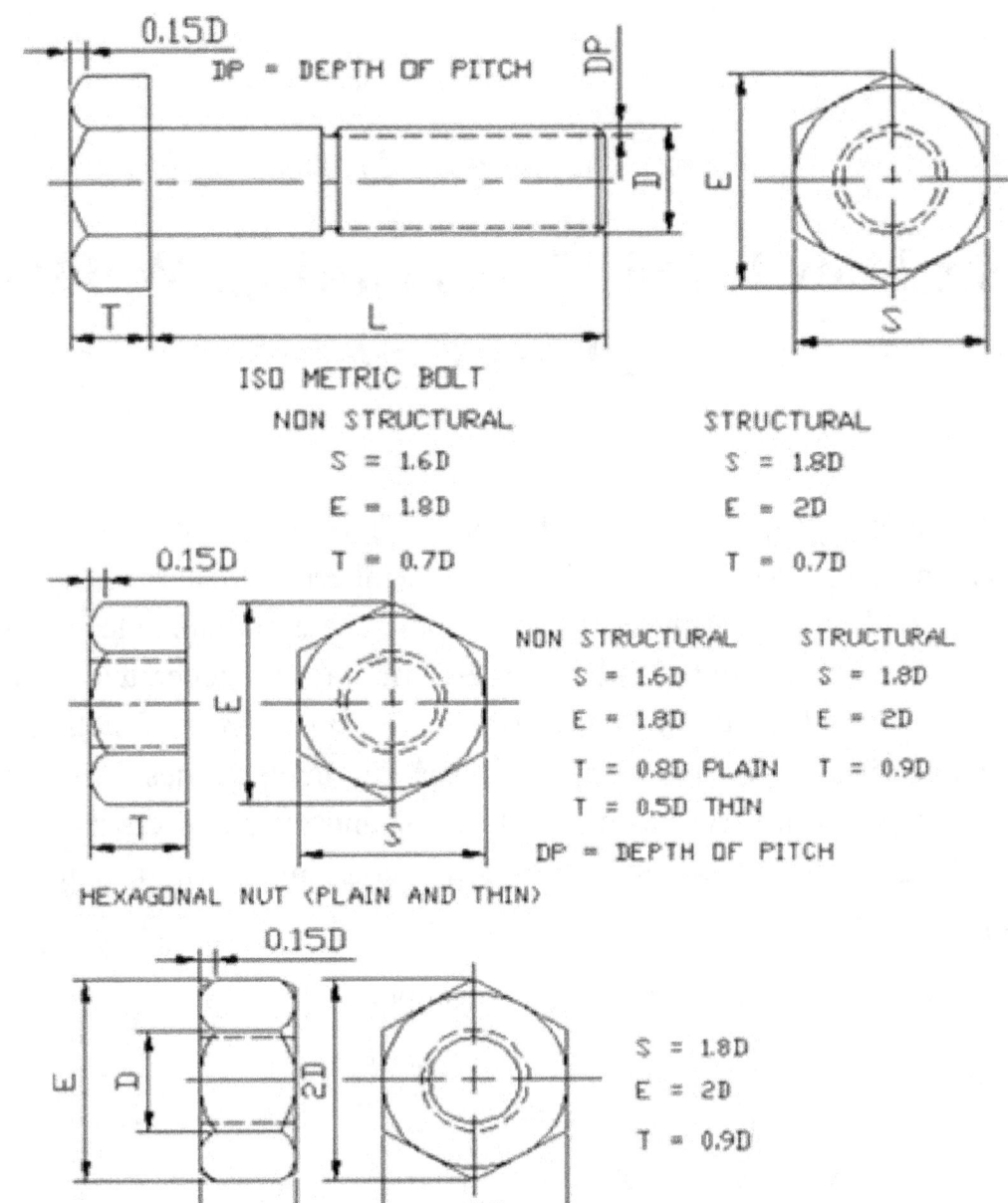

Chapter – 5

Basic knowledge of Gas cutting, Welding and brazing

Welding:

There are various processes of welding, but the most common process in use is shield metal arc welding. In this process alternative current or direct current is used from a welding power supply. The consumable electrode coated with flux is used. When the electric arc is generated between electrode and metal to be joined, electrode melts and a layer of metal is laid between two pieces to be joined. The flux coating on electrodes disintegrates giving off vapours. It serves as a shielding gas and provides a layer of slag. It protects the weld area from atmospheric contamination. It is used for welding iron and steel.

The use and selection of electrodes depends upon

- Weld material
- Welding position
- Required welding properties.

Gas cutting:

The gas cutting is also one of the important requirements of mould making skill. It must be learned for emergency need. Oxy-fuel cutting is the widely used method of common use.

Oxy-fuel cutting:

Acetylene and oxygen gas are used to preheat metal to red hot and then oxygen is used to burn away the preheated metal. The oxidation takes place at the localised place and gap is created in between two ends. It is effective on metals that easily get oxidised at this temperature. Mild steel and low alloy steels being cut easily. It is suitable to cut 1/4 inch to 12 inches thick.

When acetylene mixed with oxygen is ignited the flame temperate rises between 5700^0 F to 6300^0F. The acetylene is highly combustible when mixed with oxygen.

It can cut low to medium carbon steel and wrought iron but it can't cut High carbon steel, cast iron and Stainless steel

Apparatus used:

- Regulator
- Gas Hoses
- Non-return valve
- Check valve
- Torch

Type of torch used:

- Welding Torch
- Cutting Torch
- Rosebud torch
- Injector torch

Non-return valve:

It is fitted with hose and torch on both oxygen and fuel (acetylene) lines to prevent flame of oxygen fuel mixture being pushed back. Pushing back into either cylinder will damage the equipment or may explode the cylinder.

Check valve:

It allows gas to flow in one direction only.

Welding torch:

It is used to weld metal

Cutting torch:

It is used to cut material.

Safety:

- Welding screen with blue glass must be used to protect eyes.
- Should be careful against any leakage of fuel.
- Safety of cylinder is also very important – must check the possibility of any leakage or explosion.
- Keeping safe from any chemical exposure is also important.
- Have to be alert against any flashback while welding or gas cutting.

Soldering:

The soldering is the process of joining two or more metal items together by melting and flowing filler metal (called solder) into the joints. The filler metal used is having a lower melting point than the adjoining metal.

Brazing:

It is a metal joining process wherein filler metal is heated above melting point and distributed between two or more close-fitting parts. The torch brazing is a most common method of mechanised brazing in use.

Chapter – 6
Classification and Design criteria

Classification of wooden moulds:

- Pneumatic ramming Mould,
- Mould for Moulding with machines
- Precast moulds

The prime considerations in construction of wooden moulds are

- **Strength**:
  There should not be any deformation in mould cavity when refractories brick composition is filled and subjected to suitable pressure to get required bulk density and properties.
- **Removal of brick from mould:**
  The dismantling and assembly of mould components should be easy and less time-consuming. The brick removal should be easy without any damage.
- **Design of mould component:**
  The mould should be equipped with all necessary components. The four sides of mould cavity should provide features on four sides of brick, but when there is profile in top and bottom surface of the brick, it is to be made with some additional pressing device from top and bottom. This pressing device must be compatible with pressing equipment or pressing mechanism. This requirement is complied by lay-out.

- **Compatibility to pressing equipment:**
  The mould must maintain rigidity, shape and size when subjected to pressing load under pneumatic rammer, mechanical force or vibration.
- **Handling facilities:**
  There should be provision to accommodate crane rope for lifting during moulding operation in mould shop.

Wooden moulds Components:

- Side walls
- End walls
- Battens
- Bottom die
- Top Die
- Mounting board
- Board batten
- Nut with bolts or Tie Bolts and keys

Introduction to wooden mould:

The under given drawing of mould is simple illustration. The shape and size of mould depends upon shape and size of bricks to be made. When there is feature on top and bottom surface of brick, it is formed with dies. Mounting board is made to accompany the bottom die and top batten to top die.

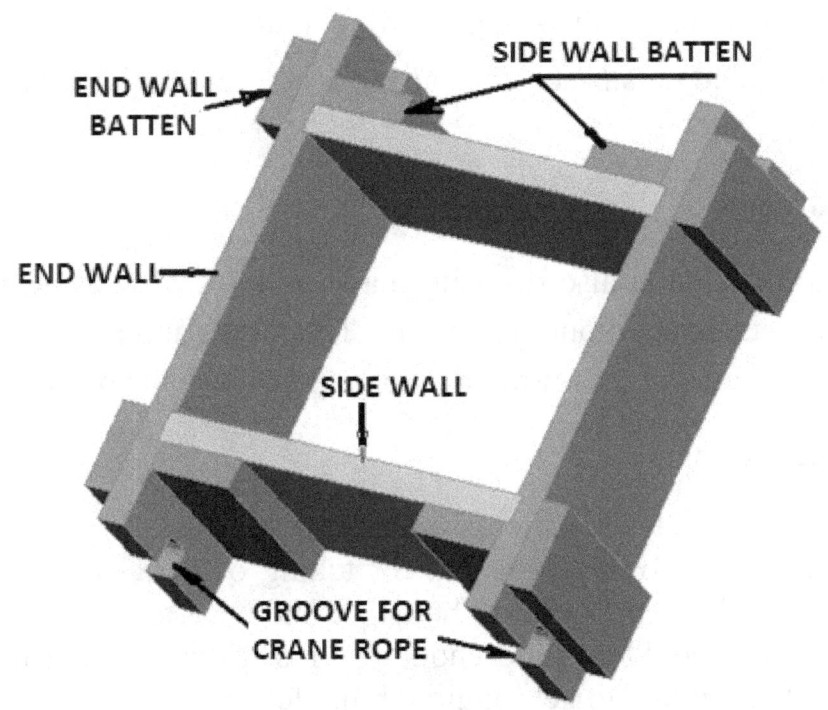

Mould making steps:

- Study drawings
- Calculate sizes considering contraction or expansion allowance
- Make brick layout on layout board
- Design mould
- Mark plate fitting area
- Design of fastener for detachable assembly
- Development of drawings for individual components of moulds.
- Note down the required sizes of wood from the layout of mould design or individual component drawings.
- Prepare templates for all necessary components' profile.

Layout:

The layout is the drawing of brick made / drawn on layout board considering contraction or expansion allowances in 1:1 scale. The layout board is made of wood or plywood. The white chalk powder mixed with glue or blue colour is applied on surface of board to increase the visibility of lines drawn on it.

Mould design:

The mould design is also drawing made in 1:1 scale considering easy, fast and rigid detachable mould assembly. The shape and size of all accessories are also decided and drawn maintaining size and shape of the internal cavity.

Design of fasteners:

The purpose of fasteners is to assemble mould accessories for making bricks and detach by removing fasteners for taking out bricks. The assembly and dismantling of fasteners should be quick and easy to reduce moulding time. The assembly should be rigid enough to resist the pressing force applied to compress the mixture to get required bulk density.

Development of individual component shape drawings

You will have the clear understanding of individual components that are to be made for mould. You will get the required sizes of wood for individual components of mould. There will not be any chance of making the wrong size.

Mould making work after preparation of Layout:

- Select the wooden planks of required length
- Clean one surface on surface planning machine.
- Clean other side and make required thickness with thickness planning machine.
- Mark the required length of planks and cut on the bandsaw or circular saw.
- Cut the battens to required width and length on the bandsaw.

- Make the width of planks equal to the height of mould by joining the planks if your single plank is not equal to the required height of mould.
- Use suitable adhesive to join the planks.
- The screws also may be used for joining the planks.
- Attach the battens with screws.
- Make drill hole in the side walls and battens to insert the tie bolts.
- Fit plates on all over inner surface of mould
- Fit plate on side wall's batten where tie bolts will be fitted

In non-availability of machines hand operated tools may be used for planning the wooden planks. The rip saw and crosscut saw may be used for sawing along the grain and across the grains of wood.

Plate fitting activity:

It is one of the important activities in wooden mould making. The accuracy in size and shape depends on the skill of mould maker. The plates are cut in pieces to facilitate bending to match and fit with the profile of wooden mould surfaces. The size of the mould cavity is made bigger by the thickness of plate that is to be fitted. It is taken care during layout of mould drawing or on detail design of mould where there is no independent design section.

Need for Plate Fitting and necessary Care:

The refractories mixture grains are highly abrasive and wood is soft. The plates are fitted on the inner surfaces of mould to prolong the intended life cycle of moulding the bricks. The plates also prevent warpage on wood surfaces and increase strength. The plates fitted in the internal cavity should not obstruct assembly of walls and loose pieces. The plates are also fitted on outer surfaces to increase strength such as on end wall battens and side wall battens to meet with some special requirements.

Chapter – 7
Practical Exercises

Pneumatic ramming mould

The Pneumatic ramming mould is made for bricks of large size that cannot be made on the press or do not have facilities to mould such large size brick on any mechanical or hydraulic press. It is also considered for making when fewer quantities of bricks are to be moulded. The pneumatic ramming mould is used for making complicated shapes but having no profile in top face, however simple profile that can be cut manually, designed for pneumatic ramming. The manual cut profile is checked with a template or profile gauge.

It consists of side walls, end walls, battens and filler block / bottom die when required. The bottom board is also associated with the bottom die. This category of mould has an internal cavity which is the negative shape of brick bounded by side walls, end walls and loose pieces matching with the features of brick shape.

Example -1: Make a pneumatic ramming mould for making fire brick of size 1250 * 600 * 450 and explain the process of moulding (Brick making). The expansion / contraction allowance is ± 0 % and the brick tolerance is ± 2mm.

Step by step explanations with the sketch is given in this example.

Step – 1 Design work:

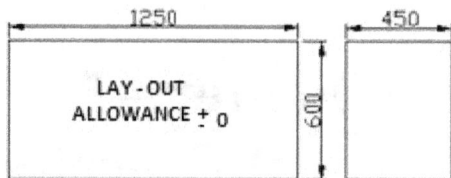

Mould design without plate:

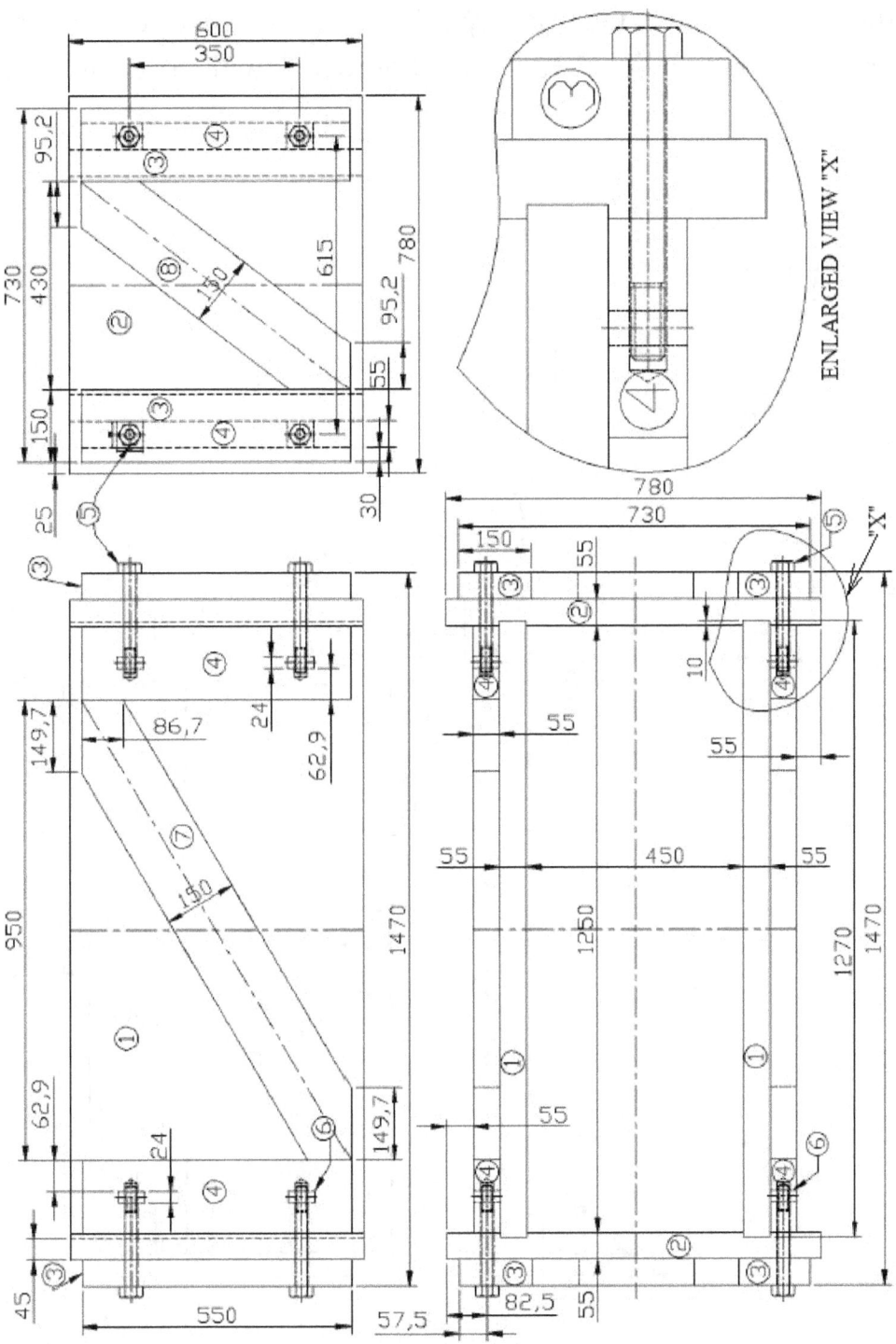

Mould design with 3 mm plate on inner walls and outer surface of battens:

Study the design; even if there is any difficulties continue reading, it will be clear. Compare the individual component drawings with assembly.

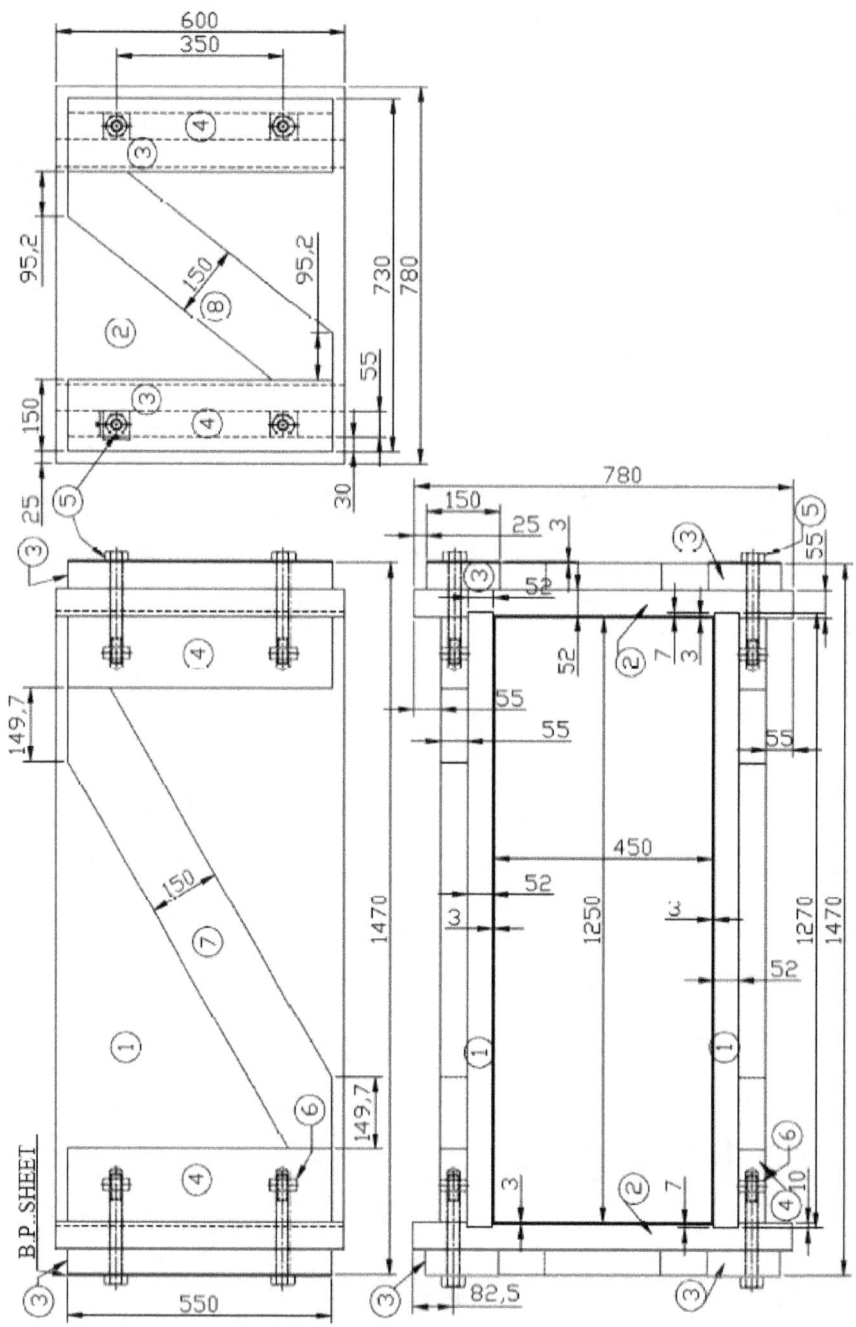

Side wall without plate: 3mm material has been reduced in thickness this drawing.

End wall drawing without plate and End wall drawing for fitting plate:

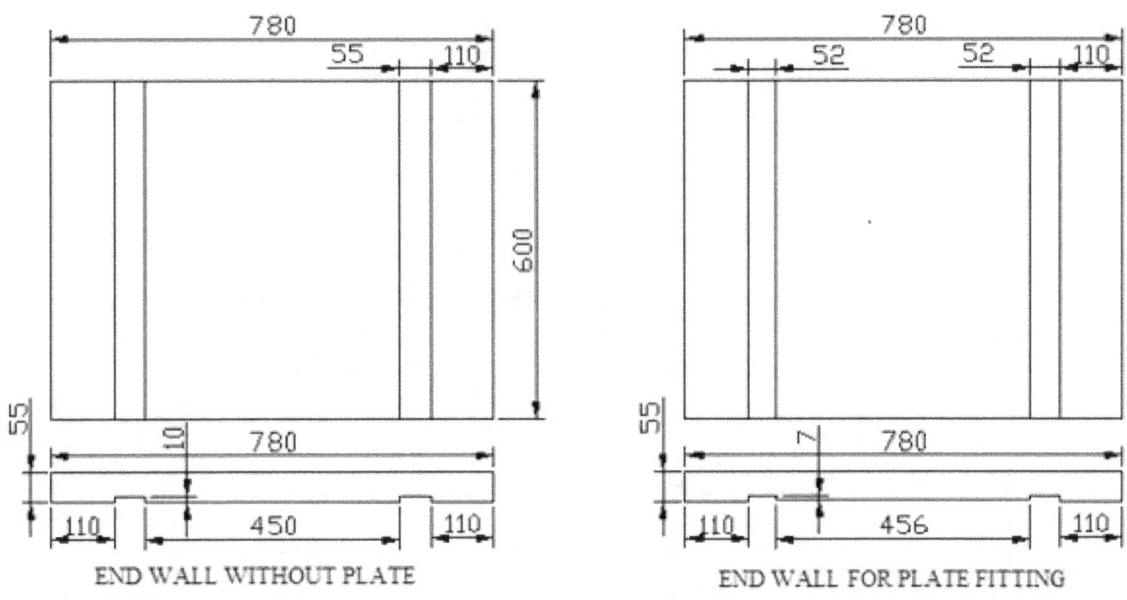

Side batten and end batten:

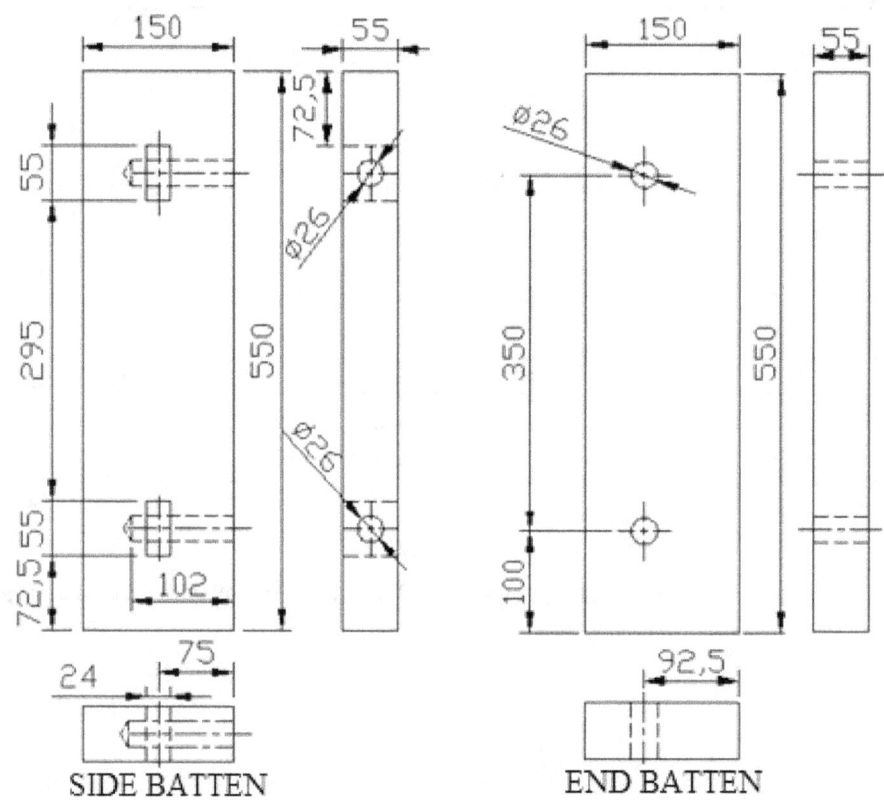

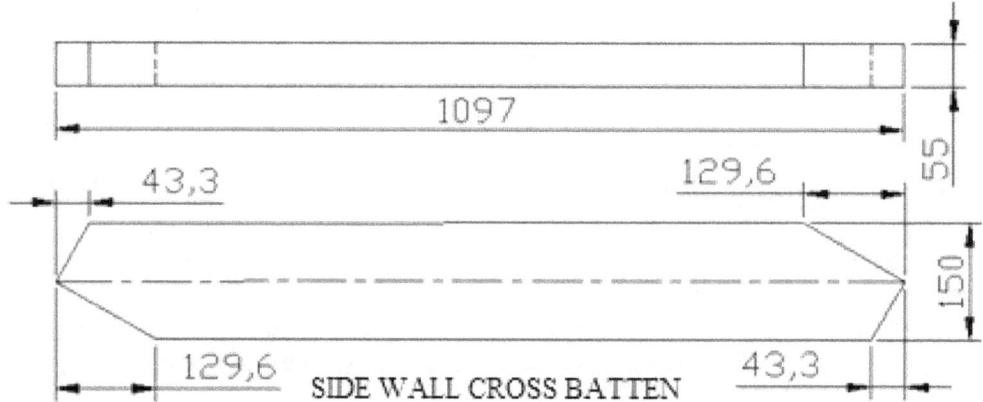

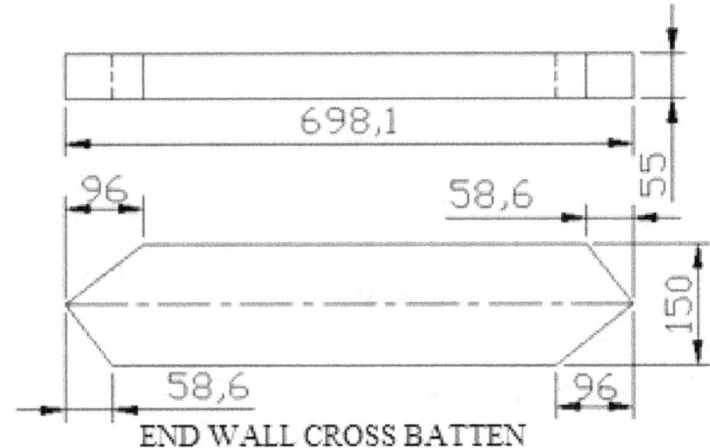
END WALL CROSS BATTEN

Fasteners for detachable assembly:

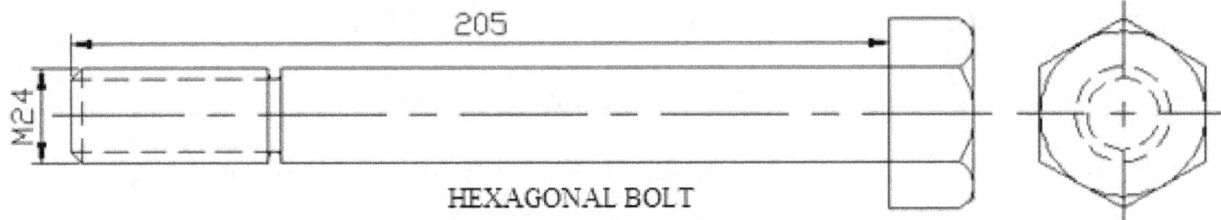
HEXAGONAL BOLT

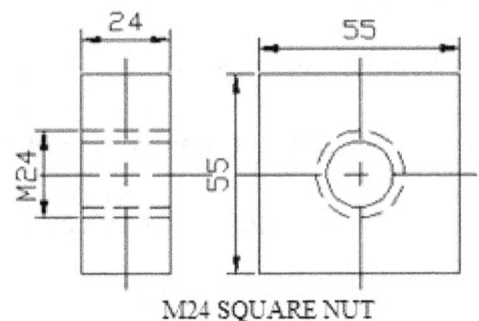
M24 SQUARE NUT

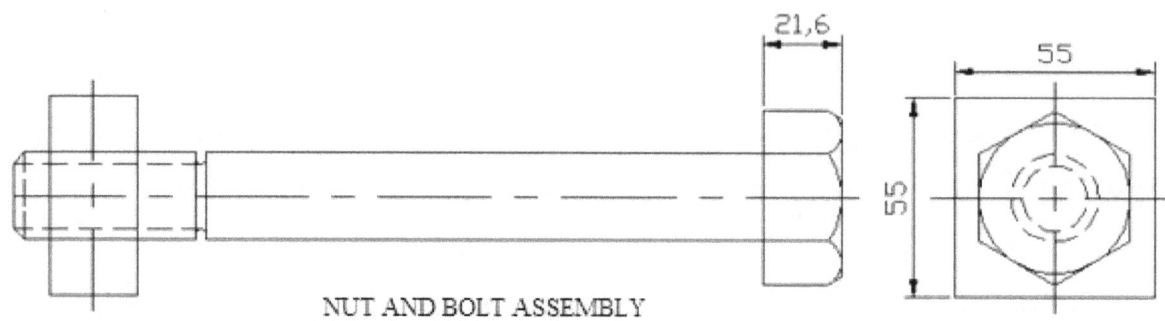

NUT AND BOLT ASSEMBLY

Step – 2 Material arrangements

Study the mould design with plate fitting and individual component drawings and find the size of wood and plates.

Item wise finished size of wood:

Item – 1 (Side wall): 1270 * 600 * 52, Quantity: 2 Numbers

Item – 2 (End wall): 780 * 600 * 55, quantity: 2 Numbers

Item – 3 (End wall batten): 600 * 150 * 55, Quantity: 4 numbers

Item – 4 (side wall batten): 600 * 150 * 55, Quantity: 4 Numbers

Item – 7 (side wall cross batten): 1097 * 150 * 55, Quantity 2 numbers

Item – 8 (End wall cross batten): 698.1 * 150 * 55, Quantity 2 numbers

Item – 5 (Hexagonal bolt) M24 * 205, Quantity 8 numbers

Item – 6 (square nut) 55 * 55 with M24 thread in centre, Quantity 8 numbers.

The sizes of 3 mm BP sheets:

- 1250 * 600 * 3 mm – 2 numbers for side walls
- 450 * 600 * 3 mm – 2 numbers for end walls
- 150 * 550 * 3 mm – 8 numbers for side and end wall battens

Wood screws: suitable Sizes and quantity needed

Step – 3 making Individual components

You need wooden planks of about 60 mm thick to prepare final thickness. The height of mould is 600 mm, join two or three pieces of planks as per availability to make the final height of side wall and end wall. Arrange wooden planks of 60 mm to have all above-mentioned items.

The wooden planks are made by cutting logs on bandsaw machines in sawmills that deal with various quality and sizes of wood. The surfaces, sides and ends have waviness that appears due to cutting teeth of bandsaw blades. The figure of such plank is given below:

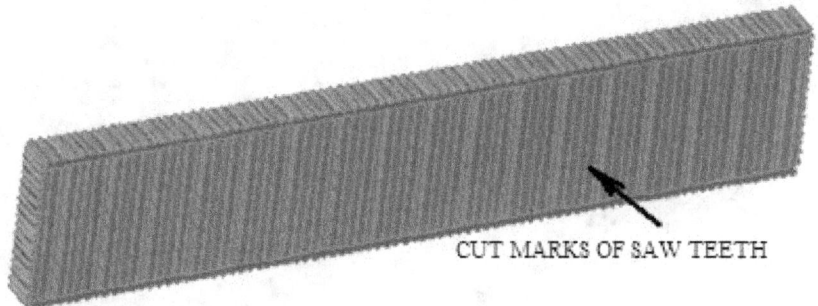

CUT MARKS OF SAW TEETH

If surface planning machine is available clean and make level one surface of planks and make one side right angle with respect to the cleaned surface. Take the planks to thickness planning machine and plane the other surfaces of all planks.

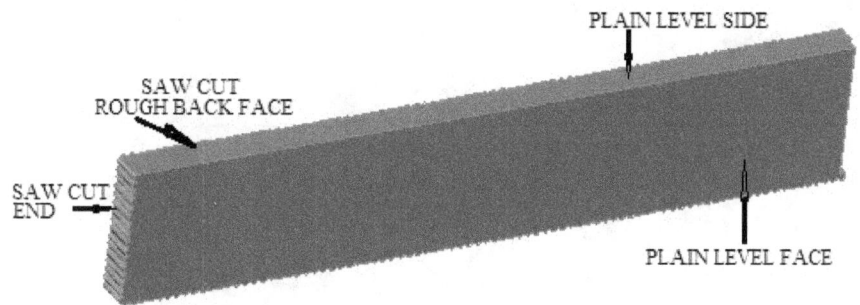

If planning machines are not available this work can be done manually also. In this case, all the required pieces are first cut on bandsaw machine or with hand operated crosscut saw or power-driven circular Saw.

The planning and levelling of face, sides and ends are done manually with hand operated planer or power driven planer putting the workpiece on the working bench or holding in carpenters vice fitted with working bench.

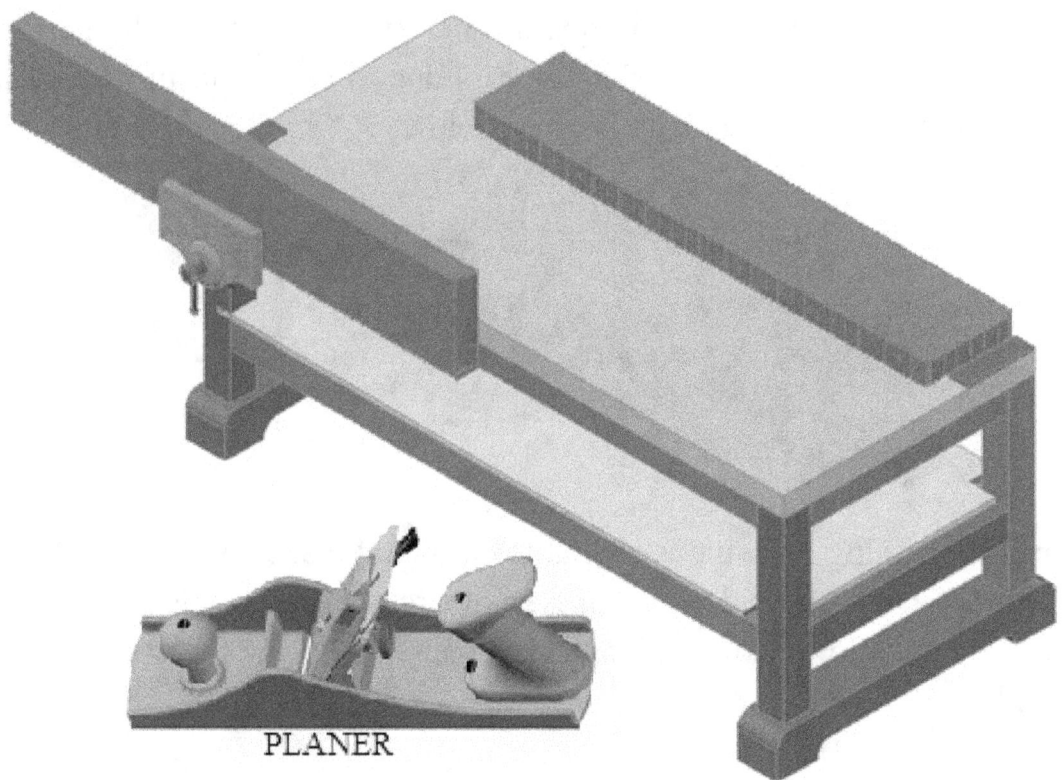

PLANER

Assembly drawing of Side wall and battens:

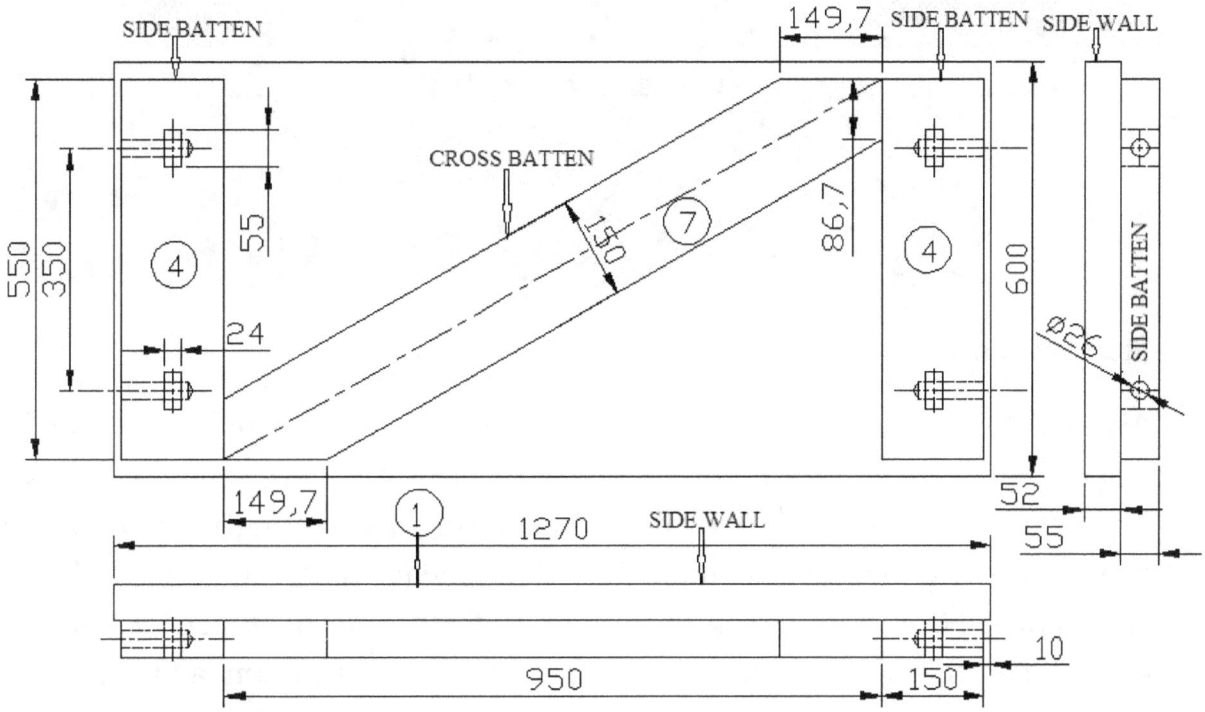

Process of making side walls:

You need two side walls with two battens for each assembly. The length of the side wall is 1270 and width is 600. The thickness is 52 mm. join two or 3 pieces of wooden planks in width as available. Cut the planks 5 to 8 mm more than 1270 with Crosscut Saw. Plain surfaces and sides, join side to side to make 600mm wide. Make V grooves in one plank and fit screws through grooves in another plank or with dowel pins (the process of joining planks has been explained in my book "carpentry work guide".

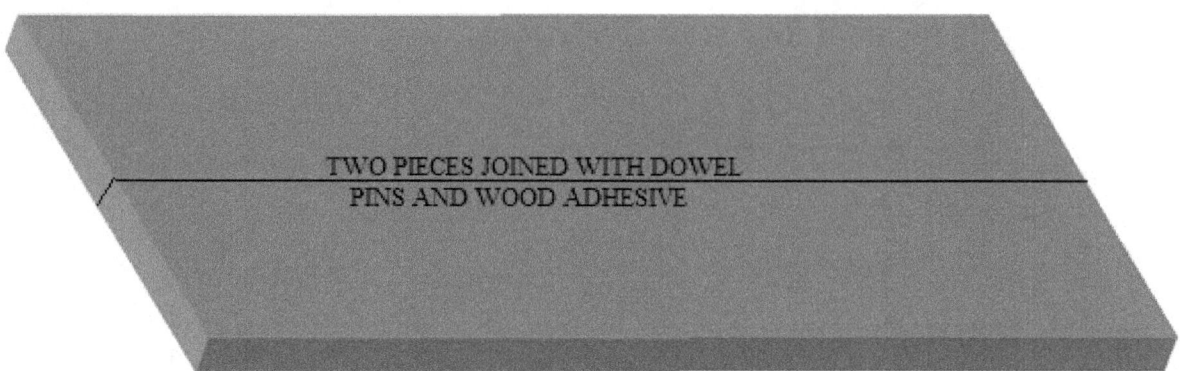

Side Battens:

Arrange 4 numbers of wooden pieces slightly more than required sizes, plain make level, warpage free and right angle all surfaces (faces, sides and ends). Sides must be right angle w.r.t face for marking and cutting special feature of the groove in correct position for fitting square nuts. The 26 Dia holes also should be in correct centre to align with taphole or threaded holes of nuts. Mark the size of slots and holes in correct position and cut with appropriate cutting tools. See the marking on batten and finished shape of slots and holes in batten in the drawing.

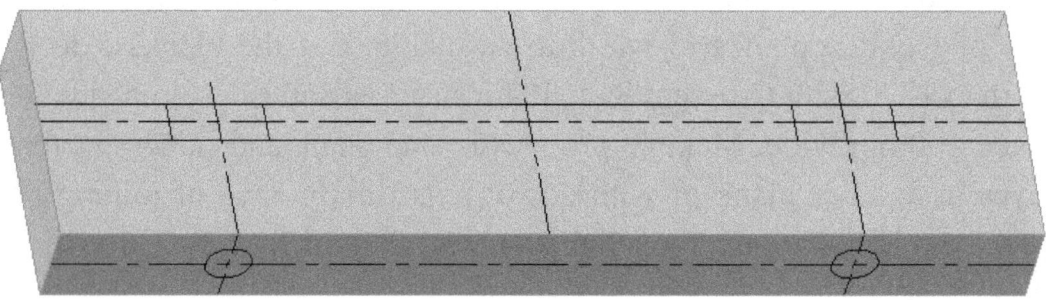

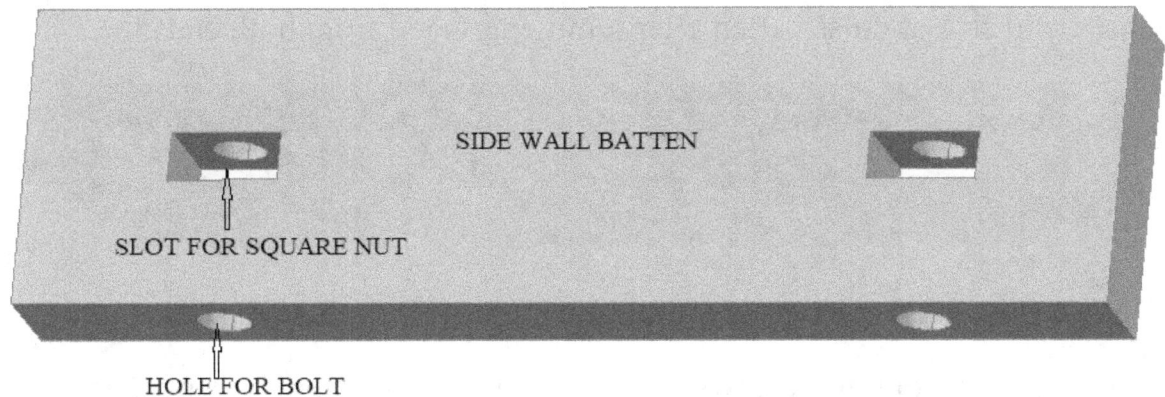

Cross battens:

Arrange two pieces of wood, cut and plane two faces and sides, to make cross battens of size 1097 X 150 X 55 for side walls.

Process of marking:

Use marking gauge for drawing centre line on workpiece. Draw the straight line a – b and c – d on side face with right angle as per layout. Join the end points on both surfaces with centre line at both ends. Mark on the opposite side of centre also taking measurement from lay out. Mark in the same way on opposite surface for reference while cutting and planning. Remove marked triangular area of wood with crosscut saw and level the surface with hand planer. Transfer all measurements from design lay – out carefully with divider.

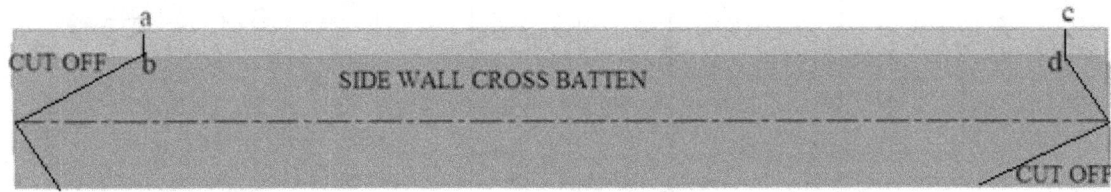

Shape and size of cross batten after removing wood from both ends:

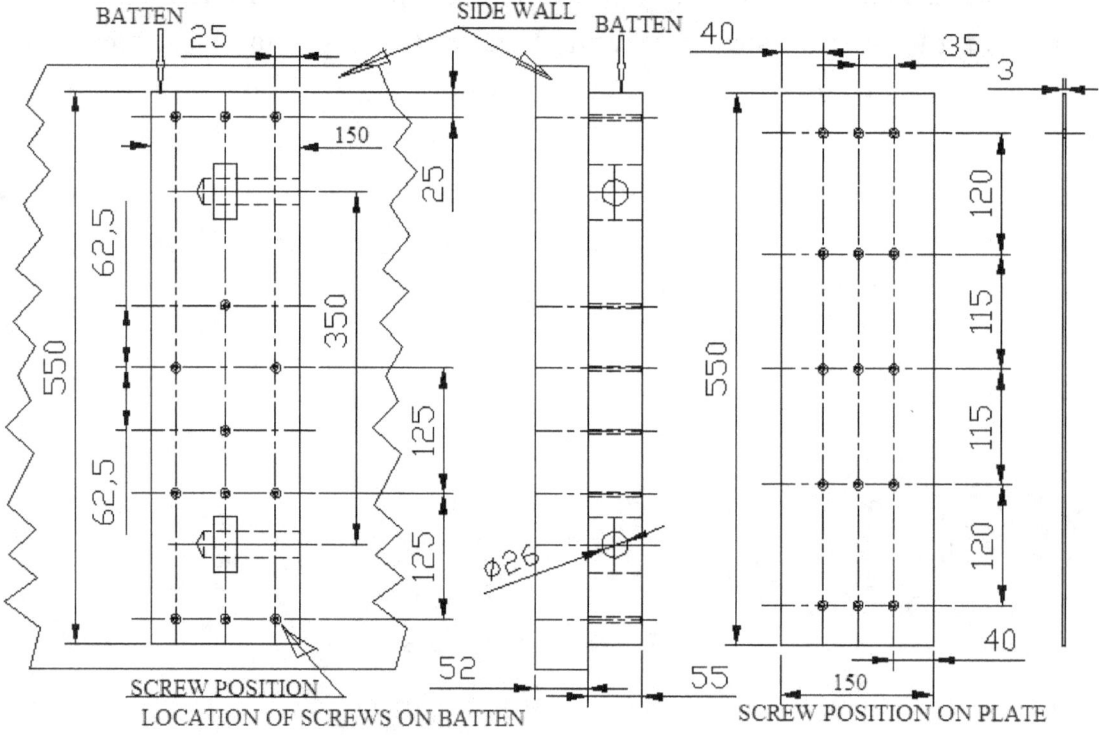

Design (layout) for screw position on side wall batten and plate:

The design layout for screws positions on battens and batten plates are essential to eliminate the possibility of overlapping of screws while fixing plate on batten. First battens are to be fixed with screws on backside of side wall then plate. The centre points of screws on batten are different than the centre point on plate. The care also has been taken not to fix screw in square nut and 26 Dia drill hole area on batten through plate. See the drawings. Fix screws of longer length that can hold plate, batten, and wall strongly.

Side wall and battens assembly

Marking location of battens on side walls:

First you should assemble the battens on both ends of side wall. The each end face of side wall will inter 10 mm in groove of end wall, side of batten will touch the surface of end wall and drill hole will match with drill holes in the end walls(see assembly drawing). Mark lines on two end of back surface of side wall with right angle. The distance of line from end face will be 10 mm. The side of batten with hole for inserting bolt will be towards end face of side wall. See the drawing, the batten "A" that will move in the direction of arrow and fix on location marked "A". Similarly "B" will be fitted at location "B". Fix these two battens with appropriate size of screw before fitting cross battens.

Assembly process of Side and Cross Batten with side wall:

Make drill holes in the battens to insert screws. The size of hole must allow the free movement of screws to be fitted with side wall. The process of attaching two wooden pieces with screw has been explained earlier. Select screw driver having matching tips with slot in screws to fix battens with wall. Tighten the screws strongly in the walls inserting through holes. The assembly should be firm, strong and rigid. Apply wood adhesive on mating surfaces before fixing screws. Place cross batten in between side battens, hold with clamp and fix screws.

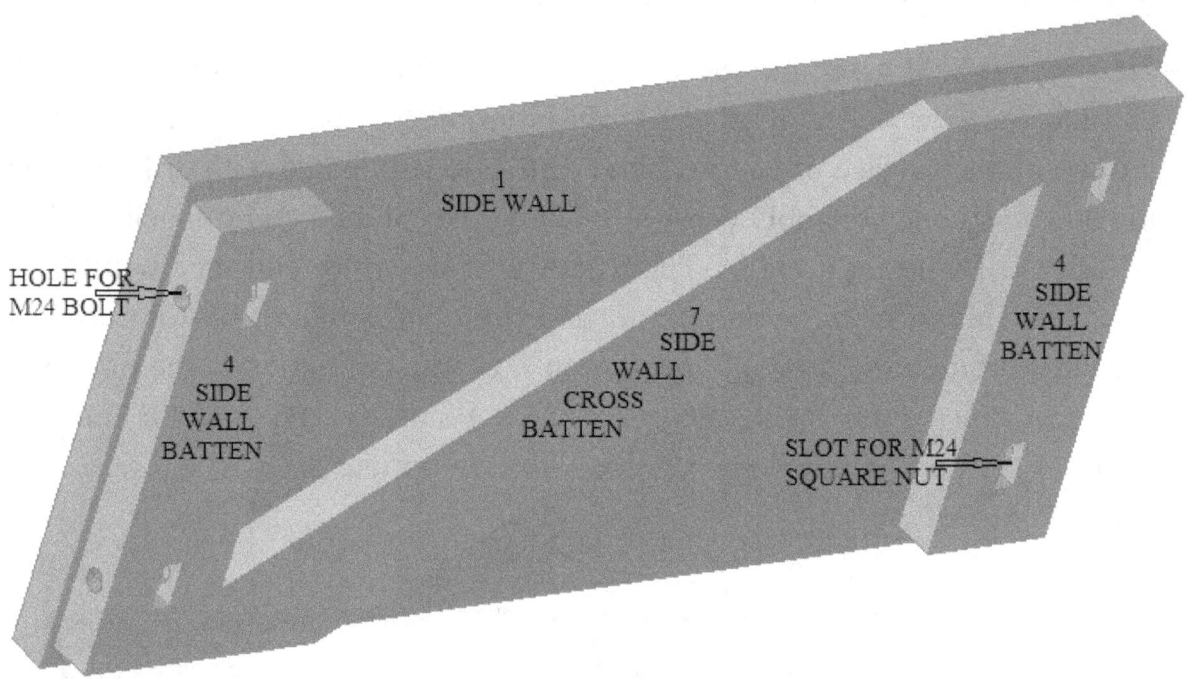

The cross section of side wall with side wall batten attached with screw:

The screws should be fitted from both sides. The screw should hold more than 70% of other part. The location and number of screws will depend on size of wall and batten but care must be taken to eliminate overlapping of screws while fixing plate on inner surface of mould cavity and on outer surface of batten.

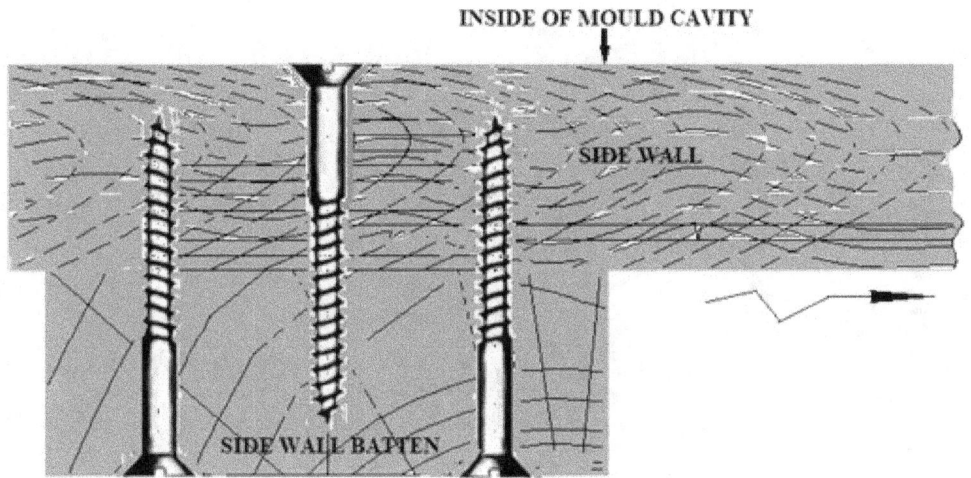

Note: It will be better to fix side wall battens with small size Allen bolts. The number of pieces will depend upon length and width of battens to accommodate the bolts. The location of bolt should not overlap the screws that will be fitted from plates.

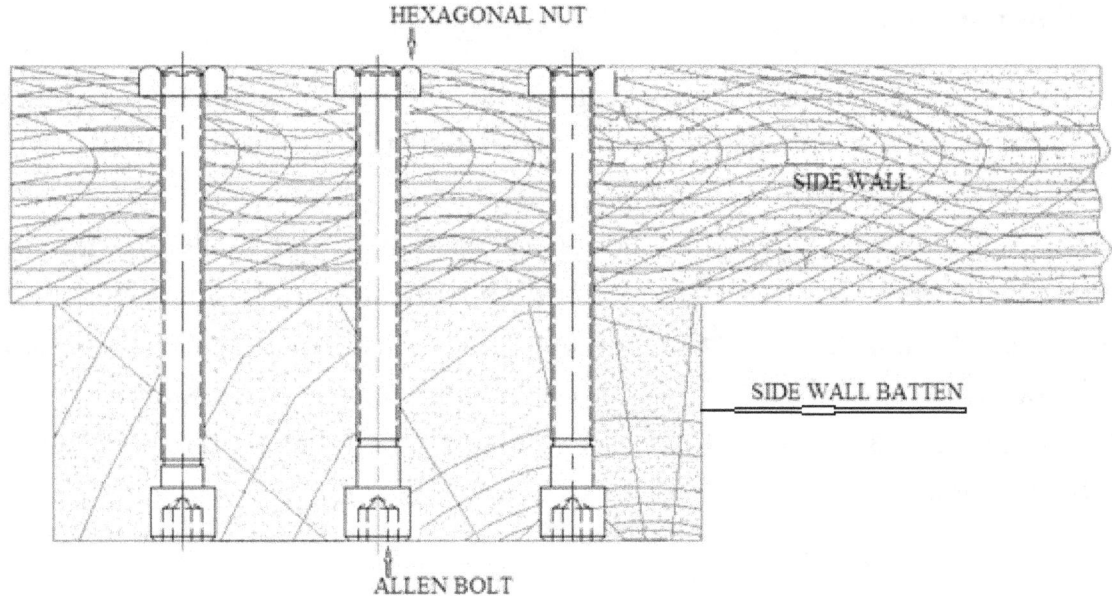

Process of fitting square nut in side walls:

Hold the nuts in hand and insert in the slots. Check the correct positioning with M24 bolt. There should not be any obstruction in fitting the bolt.

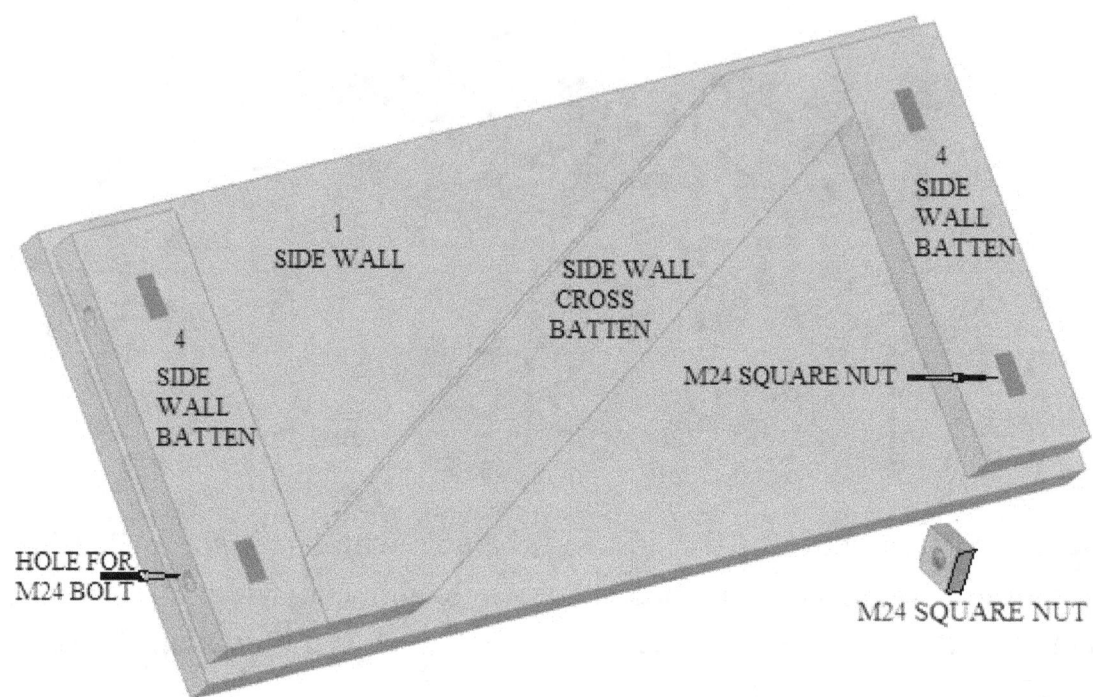

Marking on sheet:

First cut the plate 0.5 to 1.0 mm bigger in length and width than required size after marking with Right Angle and scale, file to required size. Mark horizontal and vertical lines equispaced at suitable distances from each other on the surface. Punch the intersection of each line and make countersink holes for inserting screws. The sides and ends must be right angle with each other. See the marking on side wall plate. The head of screw should flush with the surface of the plate. Only one countersink hole is shown on the plate others are also to be made on intersection of lines before fitting on surface of wood.

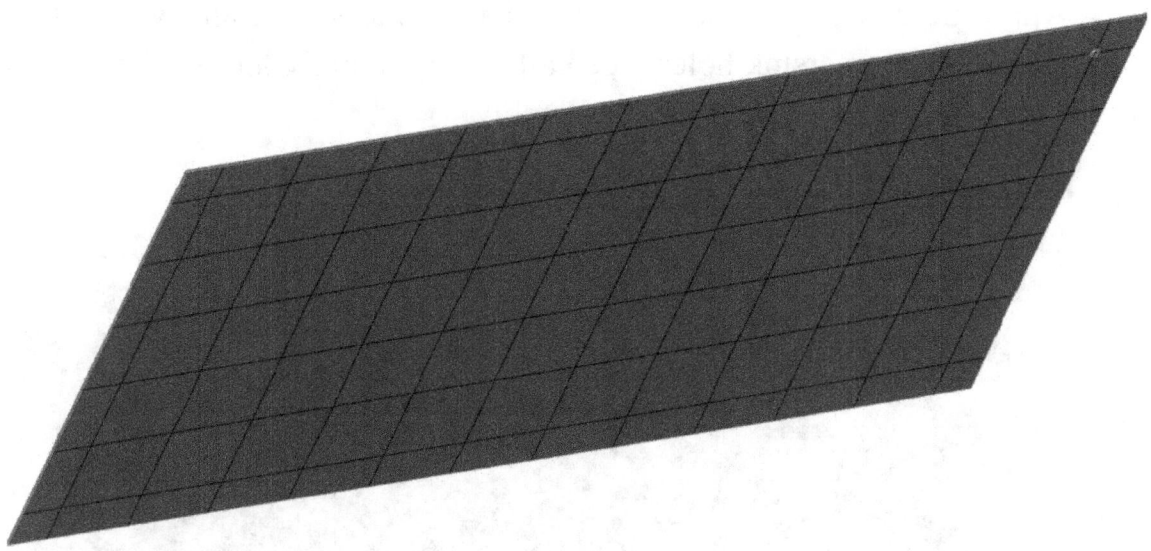

Marking on sheet for wall battens:

Follow the layout drawing for the correct positioning of holes, mark lines on surface, punch intersection points and make countersink drill holes.

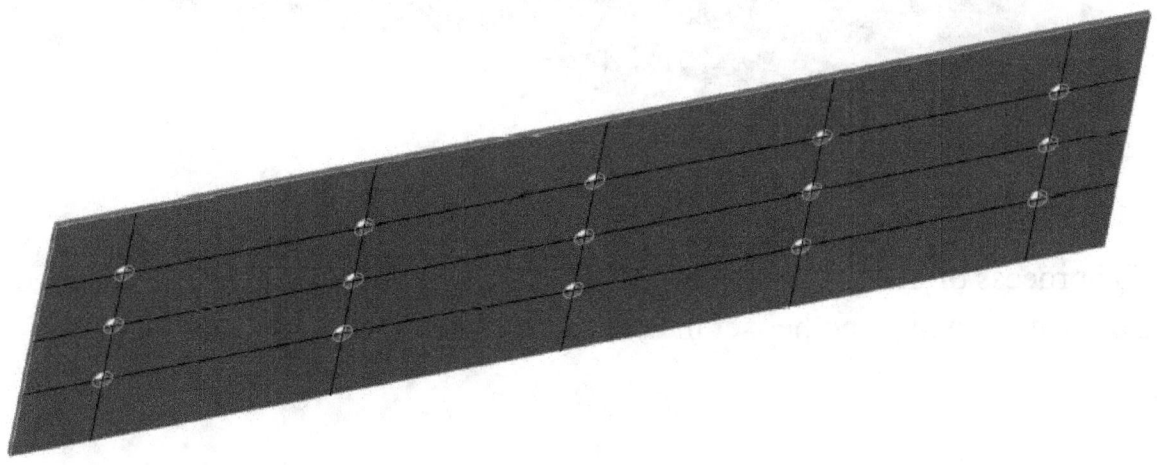

Fixing Plate on inside working face of side wall:

The 10 * 600 area on both ends of side wall will inter (fit) in the end wall groove. Mark straight line on both ends at a distance of 10 mm from end face. Align the end of plate with line and fix plate with wall by inserting screws through countersink holes Keep all screws slightly loose. The screws are to be tightened firmly after final assembly.

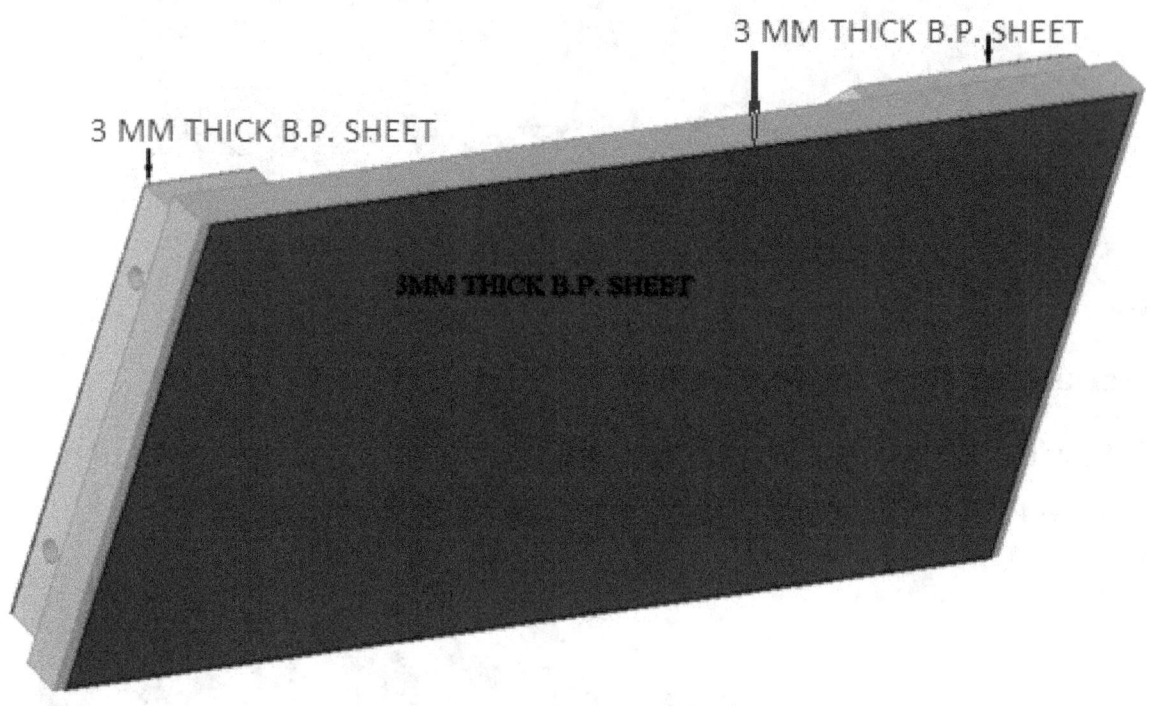

The process of making one side wall with all accessories has been explained. You need to make another set in the same process.

End wall and Battens Assembly Drawing without plate fitting:

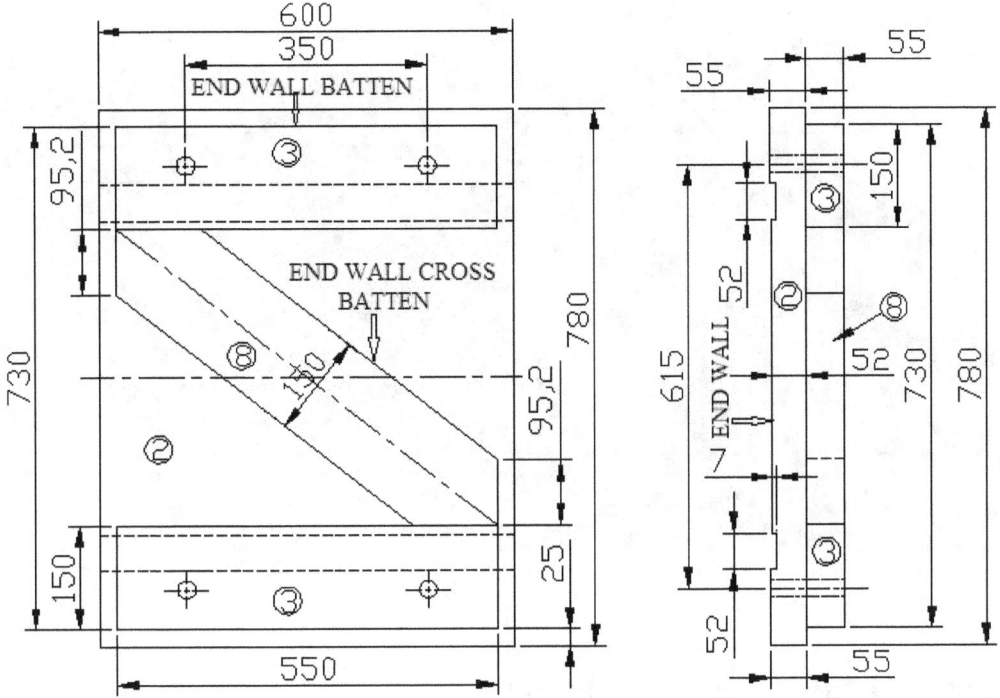

Process of making end wall:

Item – 2 (End wall): 780 * 600 * 55, quantity: 2 Numbers

Arrange wooden planks to make 2 end walls. Let us say you can get final width with 2 pieces. Arrange 2 pieces of planks and join in width with screw and wood adhesive as explained in my book "carpentry work guide". Make 2 faces, 2 sides and 2 ends plane, straight, warpage free and right angle to each other with cutting and measuring tools. Mark centre lines on face and mark lines for groove cutting and making drill holes with reference to this centre line. Use necessary appropriate tools and cut the grooves and drill the holes. Remove 3 mm thickness of wood from surface in between two grooves.

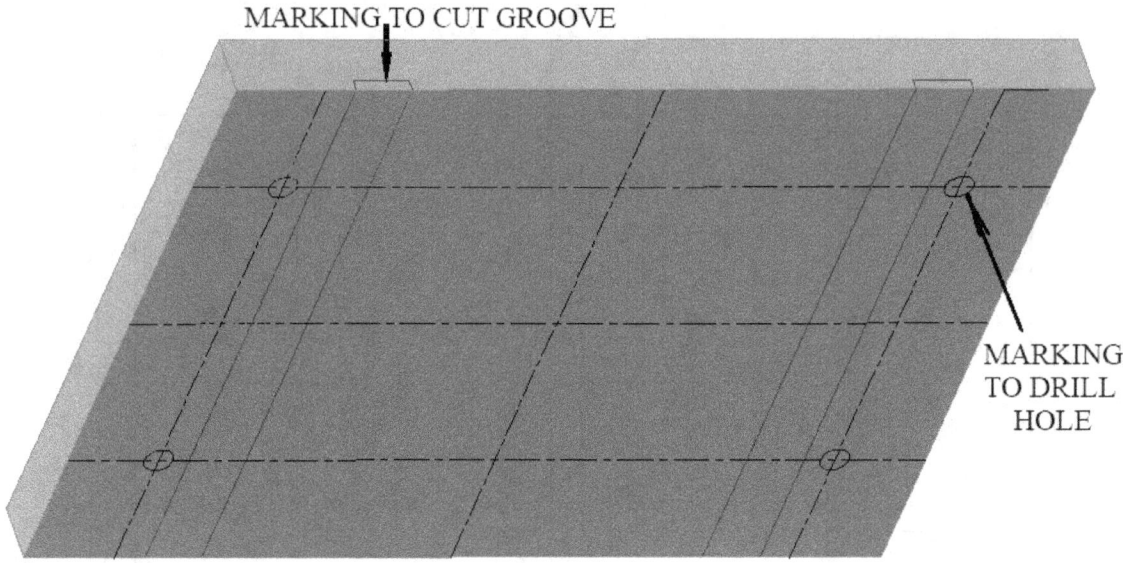

Marking and cutting end wall batten:

Arrange wooden pieces of required size, cut, plane and make right angle and warpage free surfaces, sides and ends. Mark centre lines and circle to drill holes. The centre distance of hole from one side will be = (730 – 615)/2 =57.5 and distance from centre to centre = 615 as given in drawing. Make approximately 26 Dia drill hole with drill machine.

Marking and cutting end wall cross batten:

Follow the process as explained earlier

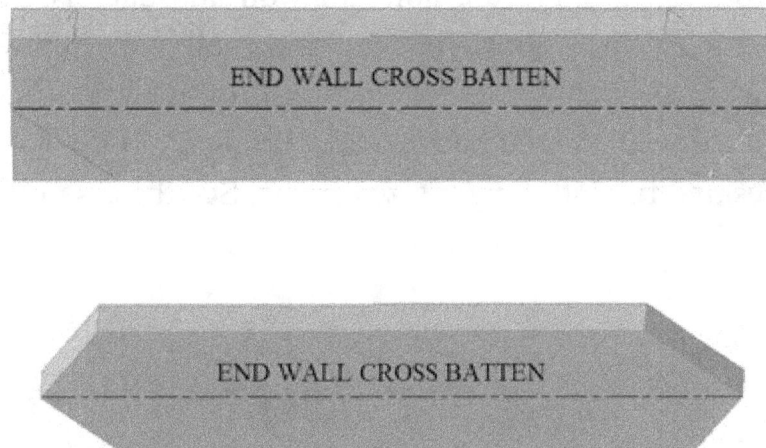

Finished shape and size of end wall and batten:

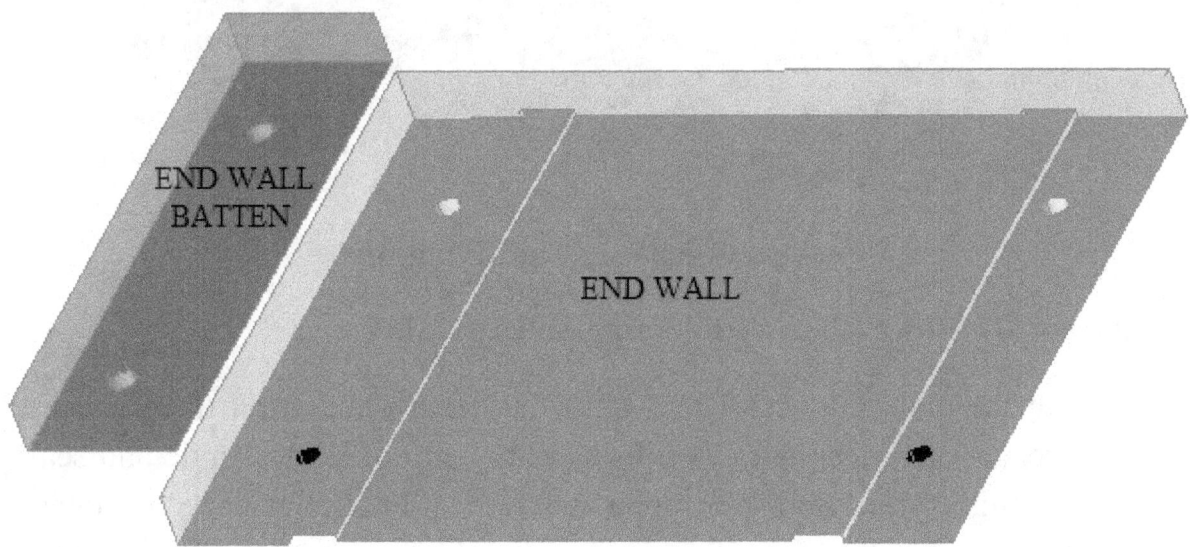

Assembly of end wall and wall batten:

Prepare 4 numbers of 26 diameter wooden plugs, insert in the drill holes of end wall and battens to align the holes of wall and battens. Hold with the carpenters clamp or vice, attach battens and wall with screws. Place the cross battens as in assembly drawing and fix it on wall with screws. Fixing of battens with end wall should be rigid and strong. Apply wood adhesive on mating surfaces and fix with longer size screws. See the assembly drawing.

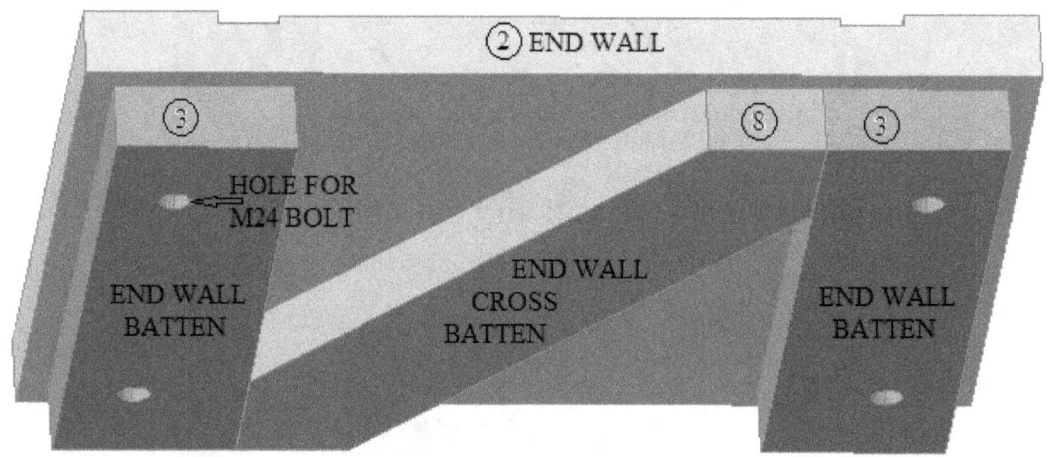

Process of making and fixing plates:

Process of cutting filing and drilling holes on Sheet:

First cut the plate 0.5 to 1.0 mm bigger in length and width than required size after marking length and width with the help of Right Angle and scale, file to required size. Mark horizontal and vertical lines equispaced at suitable distances from each other. Punch the intersection of each line and make countersink holes for inserting screws. The sides and ends must be right angle with each other. The head of screw should flush with the surface of the plate. See the marking on plates in the drawings.

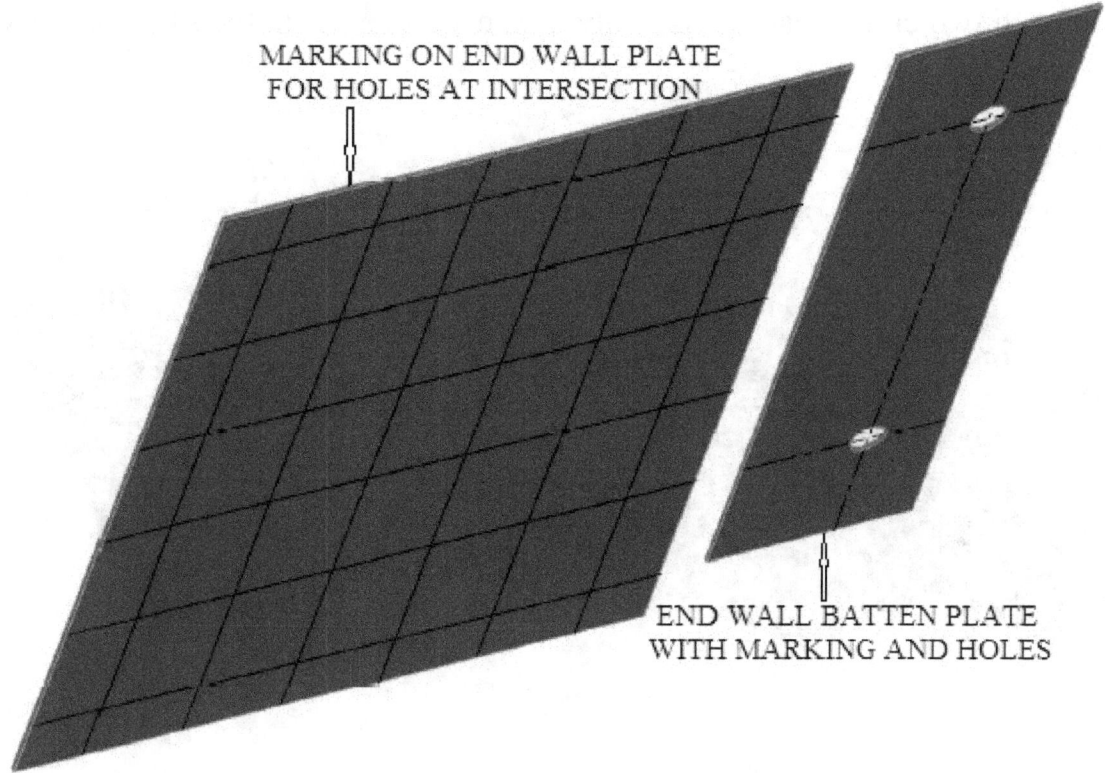

Process of fixing plates with side wall wood:

Align the end of plate with the grove lines on both sides. Hold plate and wood with C clamp and fix one screw in the centre of each end and side of plate. Now remove the clamp and fix screws in the wall through each hole made in plate. The screws must be strongly tightened by screw driver and there should not be any warpage on the surface.

Process of fixing plate on the end wall batten:

Place end wall plate on the surface of wall battens in alignment with holes, insert two wooden plugs to align. The face of plate with countersink will be outside and plane surface with holes will seat on surface of batten. Hold wall, batten and plate with clamp and fix at least two screws then remove the clamp. Now fix screws firmly through each countersink hole. Similarly fix plate on other side batten.

You need two end walls for assembly, make another set in the same process,

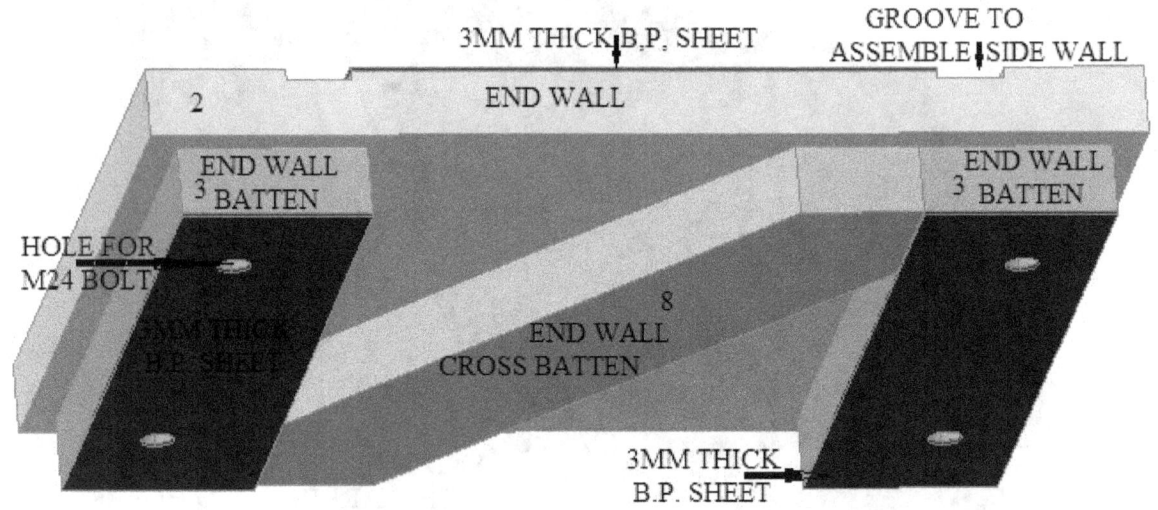

Assembly of side wall and end walls with fasteners

Place all components of the mould subassembly on the working bench. Insert two side walls end face in the grooves of one end wall. Align another end wall's grooves and push to insert end faces of side walls in grooves. Insert bots in the holes and rotate clock wise. You had left side wall plate loose. Now tighten the all 8 numbers of hexagonal bolts in the nuts fixed with side wall battens. This action will locate the side wall plates in proper position if there is any misalignment. Dismantle the mould and tight the all screws with screw driver as best as possible. Assemble the mould again and send it for moulding after checking the sizes

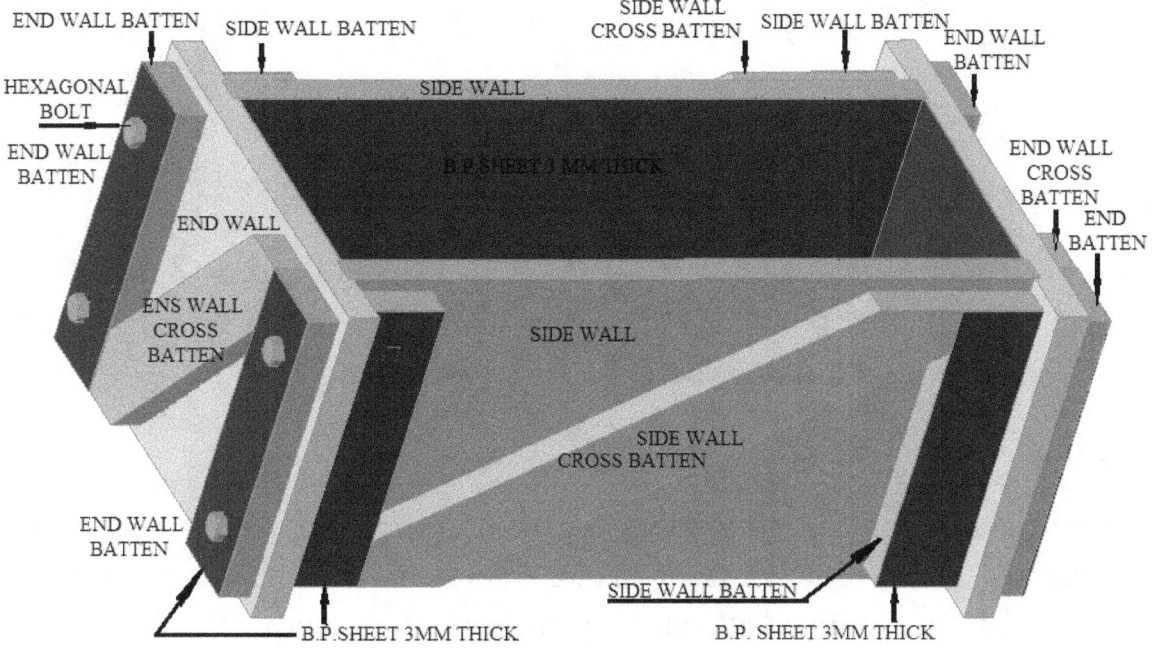

Top cover frame:

The top cover is very important to save the top surface of mould from damage while ramming mixture with pneumatic rammer. Arrange wood and make as in drawing. The joint visible in 3Ddrawing is half lap joint.

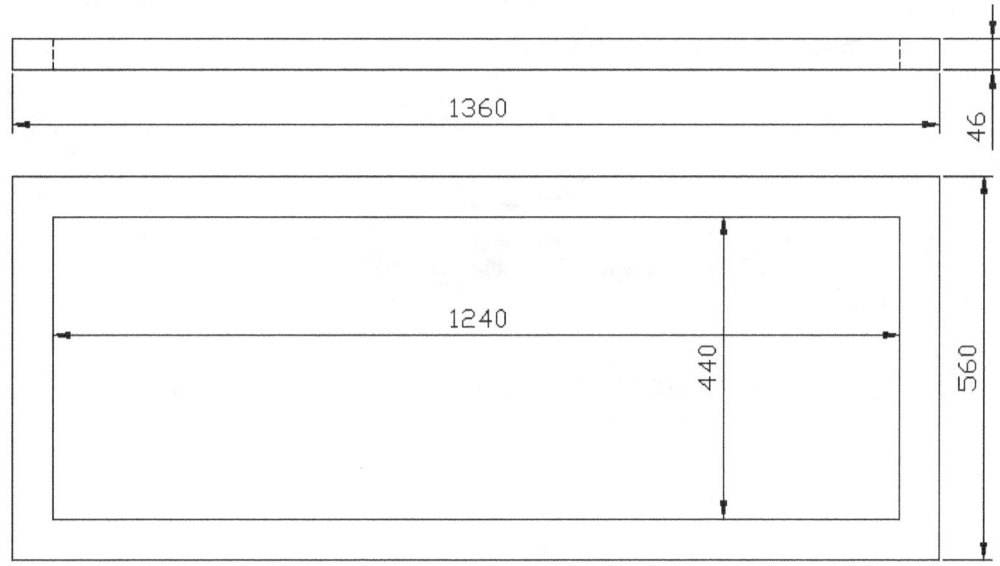

Moulding bricks with brick mixture

Pneumatic rammer:

The pneumatic rammer is working on the principle of pascal's law. The compressed air is used as fluid in the pneumatic rammer. It consists of cylinder, piston, hammer attached to piston and a returnable valve. The pneumatic rammer is connected with pipe to tank which is source of compressed air. The hammer works on the principle of differential air pressure supplied through returnable valves upon and under the piston to push the hammer. The valve works as inlet and outlet at fixed interval of time. This system sets it into reciprocating motion. The hammer goes up and down at regular interval of time. The downward movement of hammer with greater force used to compress the brick sand mixture.

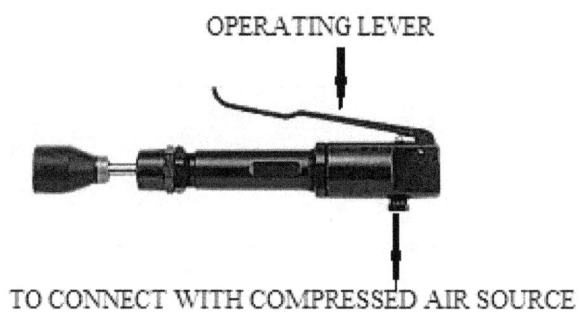

The process of mould making:

Place the mould assembly on moulding table. Place top cover on the upper surface of mould assembly. Align the mould cavity with top cover cavity. All the plates inside the cavity must be under the top cover to protect from the impact of rammer while pressing the mixture. The movement of pneumatic rammer is under manual control, there is a possibility of damage if plates are not covered by top cover. Clamp the top cover in position as illustrated in drawing with available mechanism in moulding shop. Prepare the brick mixture to have required Bulk density and quality to suit the customer's need.

Weigh the mixture on balance to get required bulk density. Compress the mixture in the cavity of mould with pneumatic rammer. Pack the mixture uniformly in the entire cavity under uniform pressure to have uniform density in the brick. Ram extra mixture in the cavity above surface of mould in the top cover. The pressing of Mixture should be from bottom to top gradually with uniform pressure after putting mixture layer by layer at regular interval at different levels.

Loosen the clamp remove the top cover, cut off the extra mixture from the mould surface and polish the surface of brick, Take off the mould with brick from moulding table and place on pallet that can accommodate it. Dismantle the mould and remove all parts leaving brick on pallet. Reassemble the mould for next operation. Send the green brick for further processing.

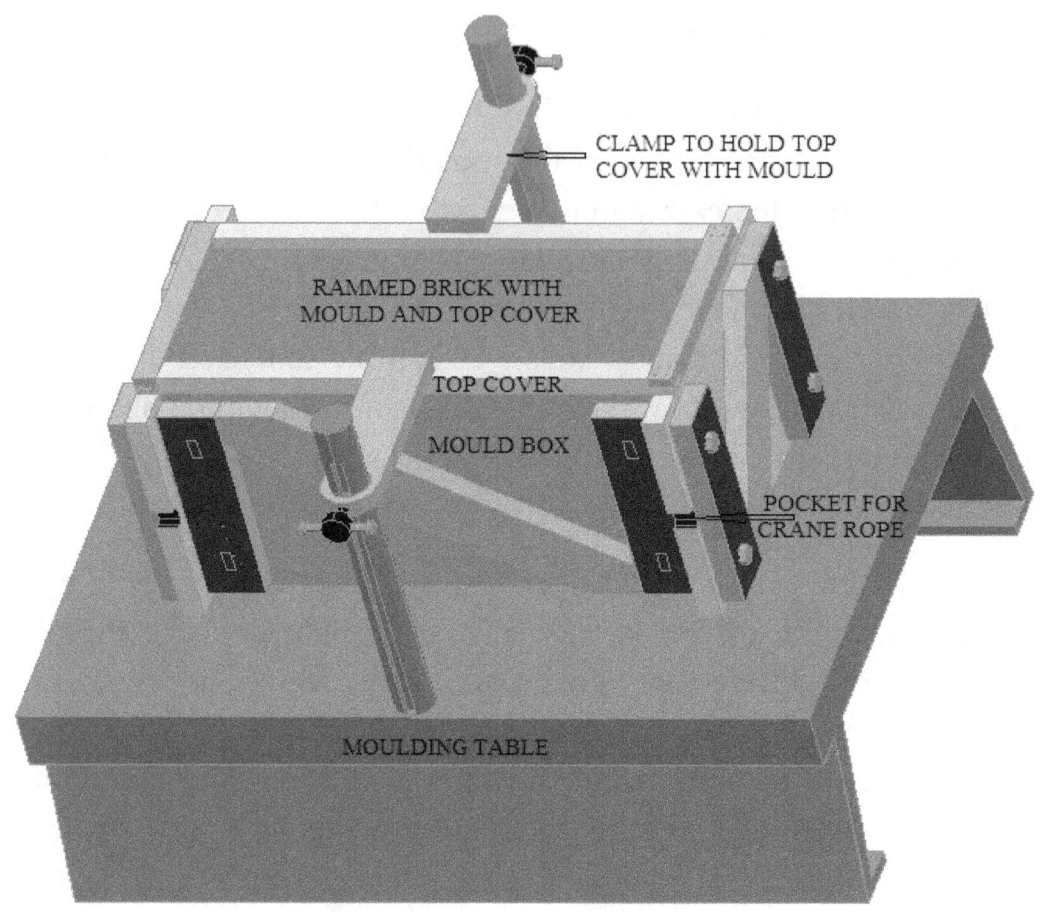

Mould taken out from moulding table with crane:

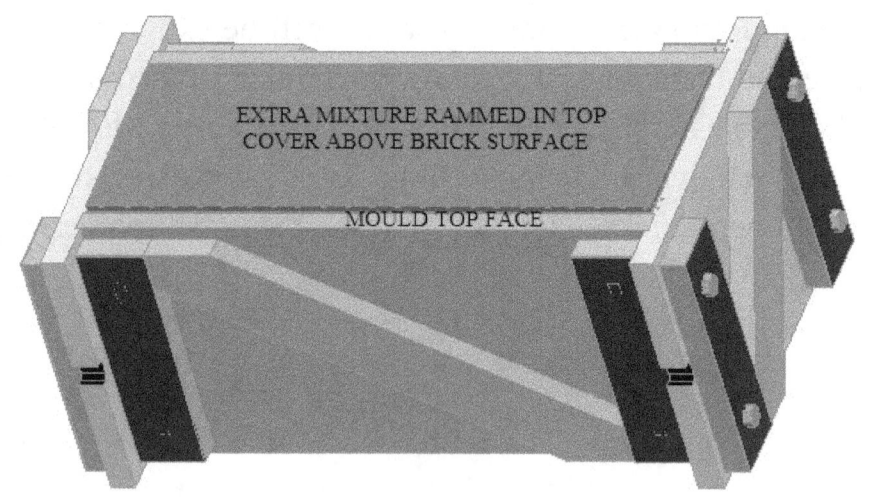

Brick surface polished after cutting of extra rammed mixture:

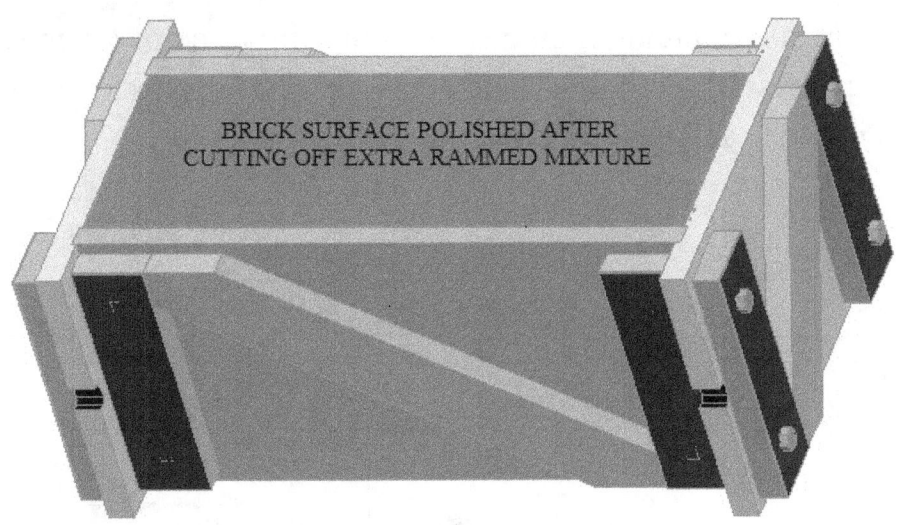

Dismantling process of mould to take out brick:

Place the mould with brick on suitable size of pallet. Hit on two ends of one end wall's inner surface gently with wooden mallet after removing all hexagonal bolts and remove the end wall. Similarly remove opposite side's end wall. Pull away two side walls also from the surface of brick. Send the brick to drawer after polishing if required.

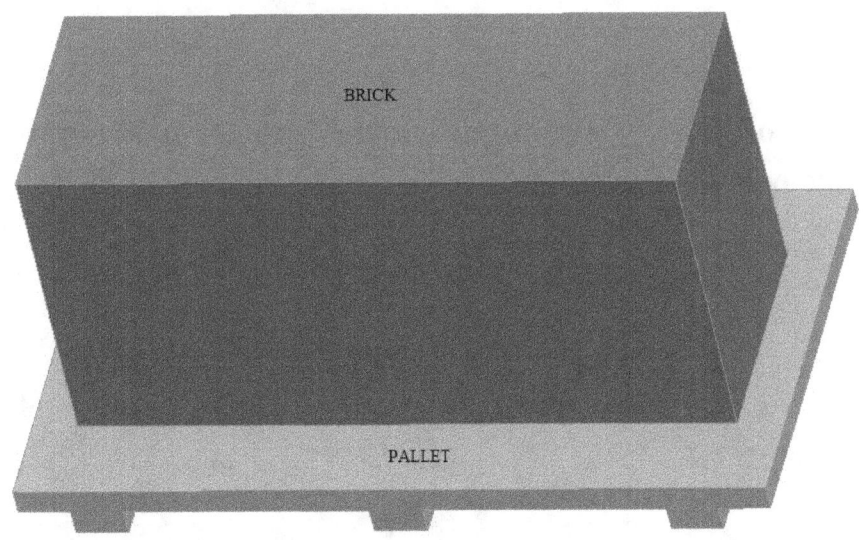

Example -2:

Make a pneumatic ramming mould for making fire brick of given shape and explain the process of moulding (Brick making). The expansion / contraction allowance is ± 0 % and the brick tolerance is ± 2 mm.

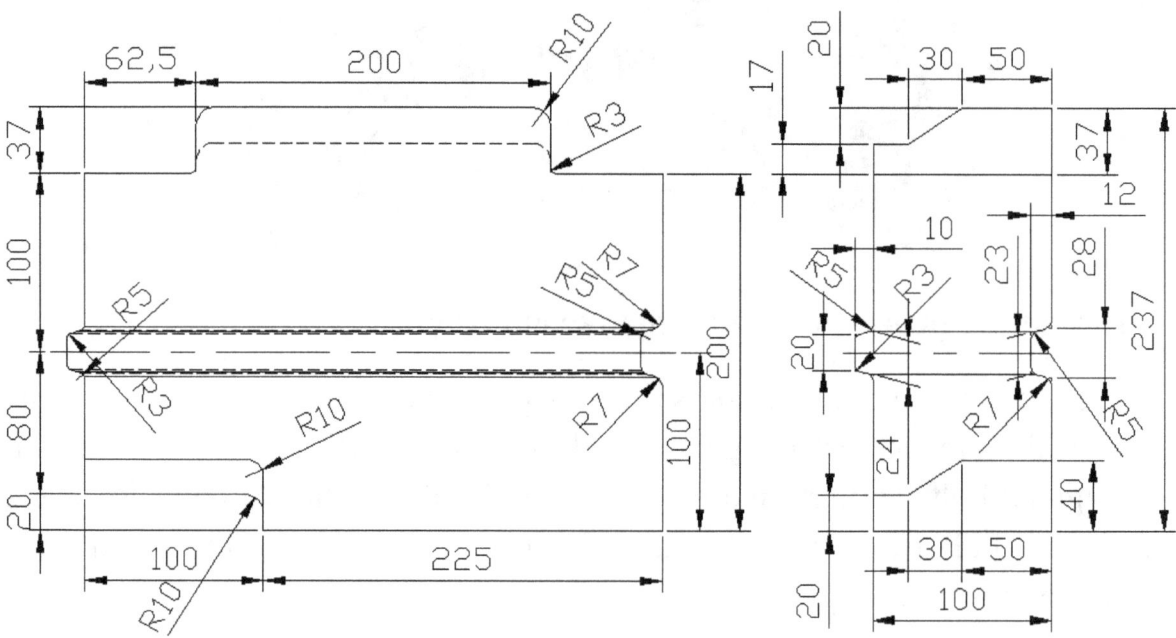

Pictorial view of brick with tongue on the upper surface and undercut profile at one corner is in the front:

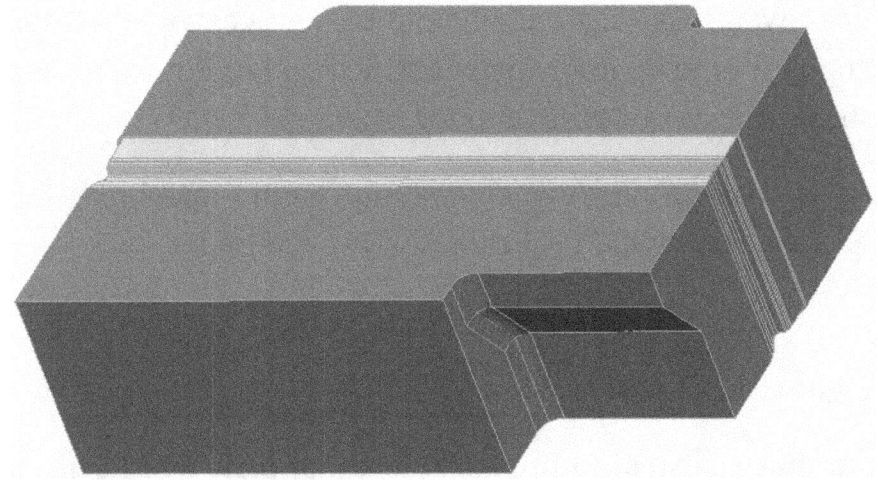

Pictorial view of brick with groove on the upper surface and back side projected profile of previous drawing is drawn in the front:

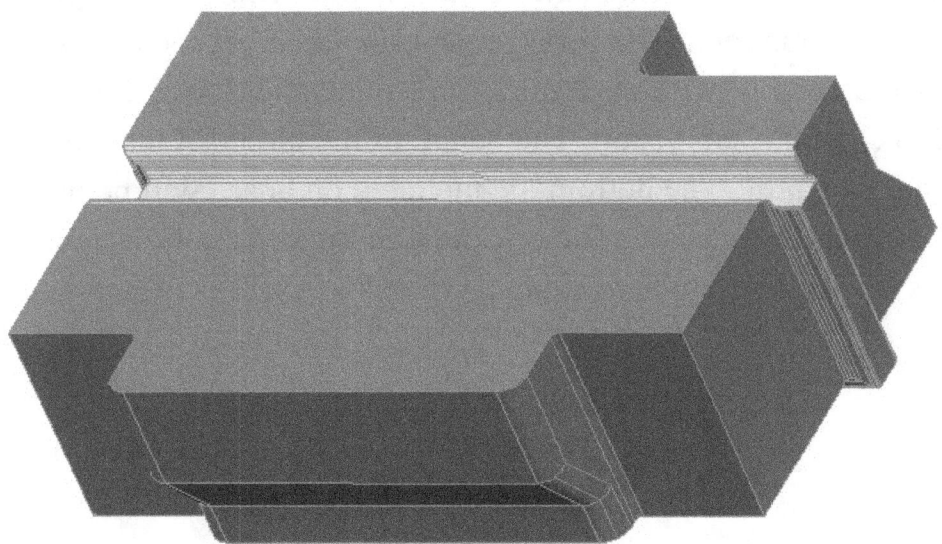

A practical way of making Pneumatic ramming mould for above shape brick

The above drawn shape can't be ejected from fixed mould on press. The features on four sides can be formed with wooden mould having negative shape on four walls and feature of bottom surface with bottom die. The feature of top face is to be made manually with a top die with tongue profile (negative shape and size of groove in the brick).

The process of making pneumatic ramming mould and brick formation is explained Step by step with the sketches in this example.

Step – 1 Design work:

Mould design without plate:

The internal cavity is equal to brick size without plate and height is equal to brick thickness + bottom die thickness. The mould drawing will be without any increase or decrease in dimensions as there is no expansion or contraction in brick after firing. The mould cavity sizes will be same as brick sizes. The design drawing comprises of internal cavity same as negative shape and size of brick and outer shape with sufficient strength to resist impact load of rammer. The arrangement also has been made to assemble and detach assembly with ease. Few firm lines and dotted lines have been omitted to eliminate confusion. The height of the mould is increased equal to bottom die thickness. The bottom die mounted on board will be placed in the mould cavity at bottom side. The thickness of bottom die is 47 mm and brick thickness is 100. The height of mould will be 47 + 100 = 147 mm.

Front and Plan View of assembly drawing without B. P. Sheet and tie bolts:

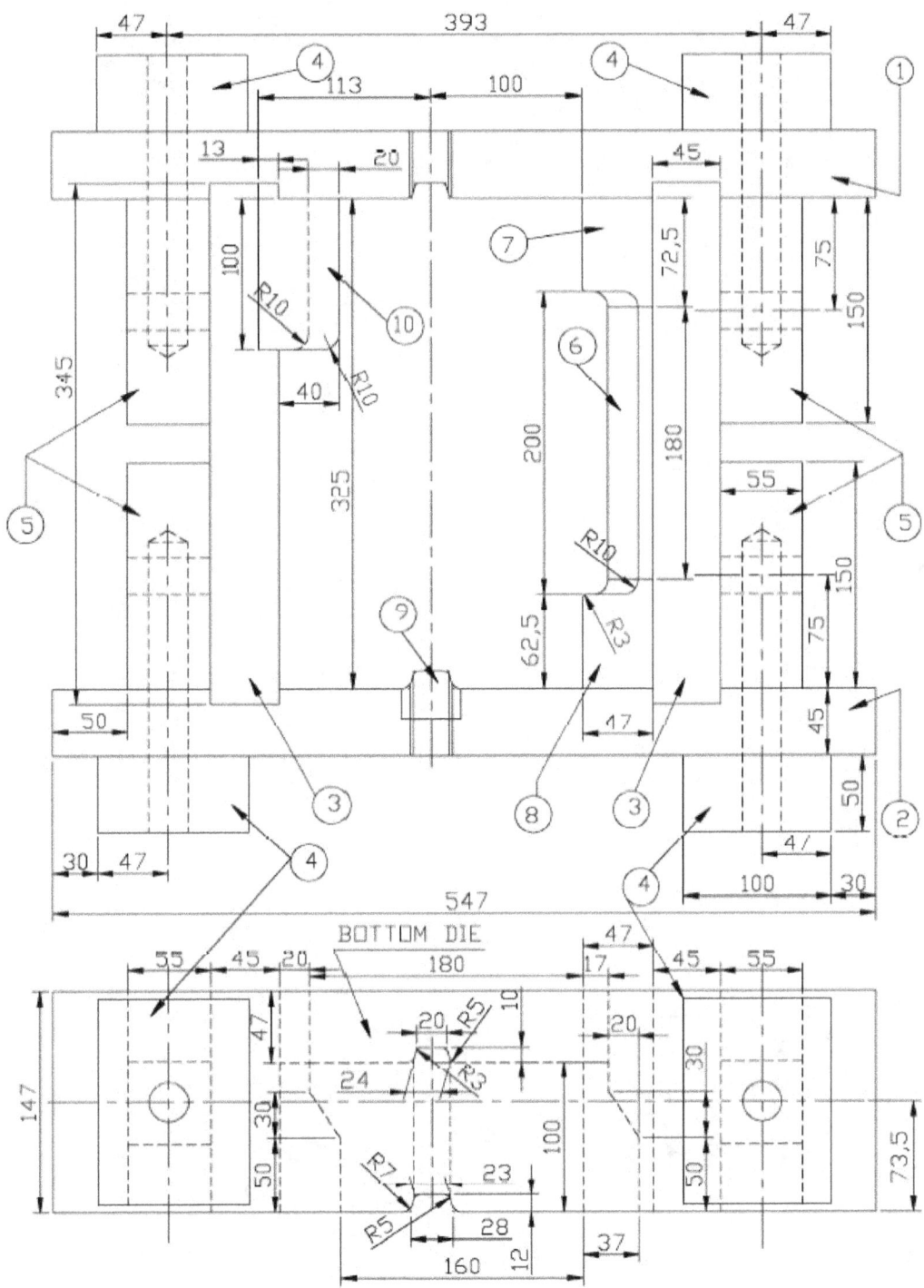

Mould design with marking for plate fitting:

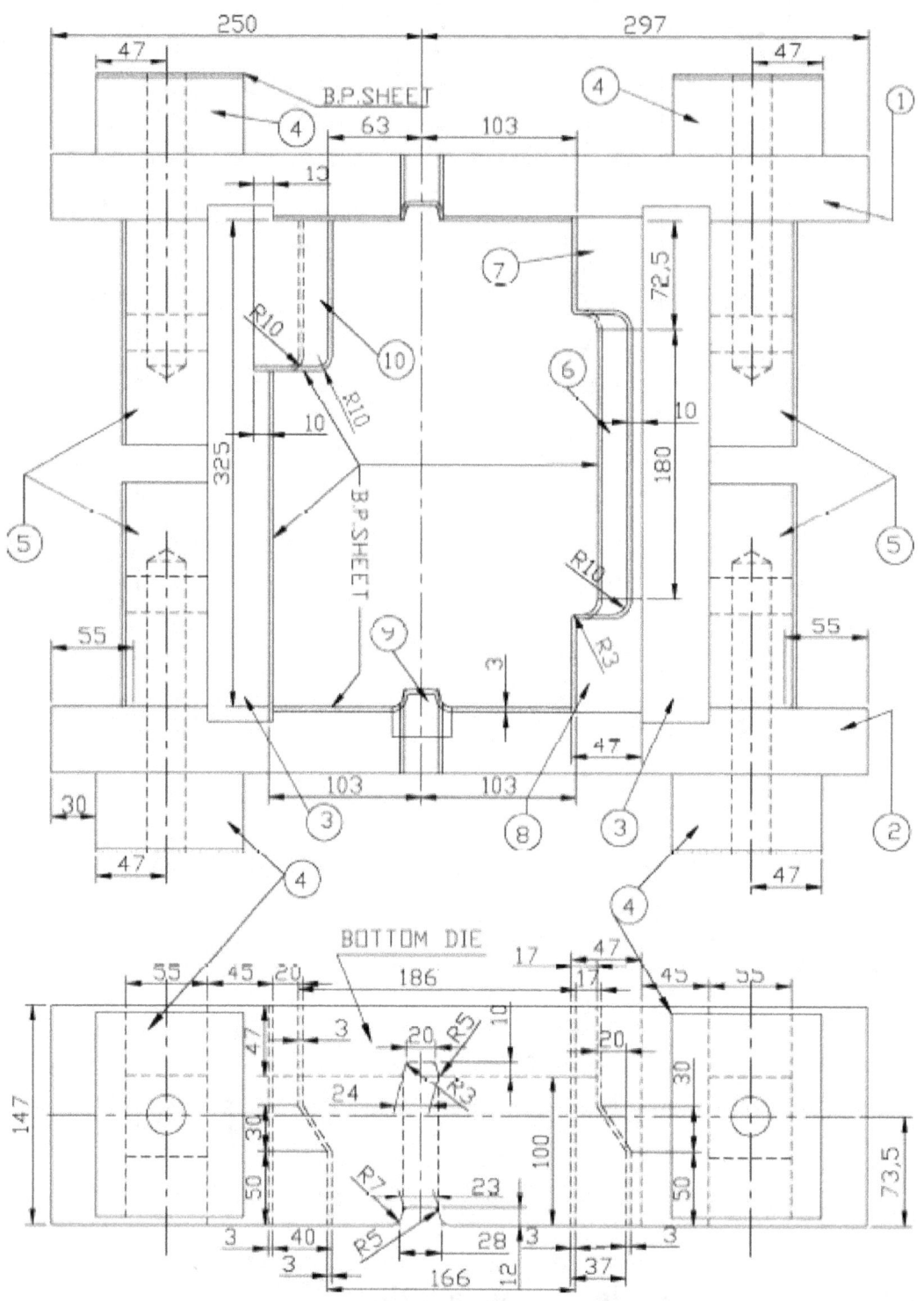

Assembly drawing of item – 1 and item – 4 with space for fitting plates:

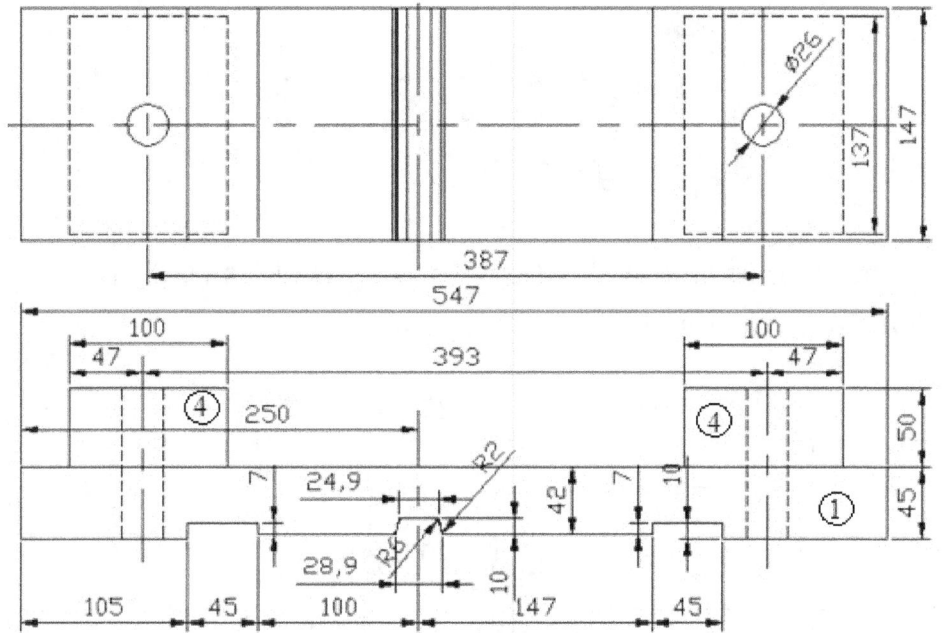

End liner (Item No – 2), End battens (Item no – 4) and loose piece (Item – 9) with space for fitting plates:

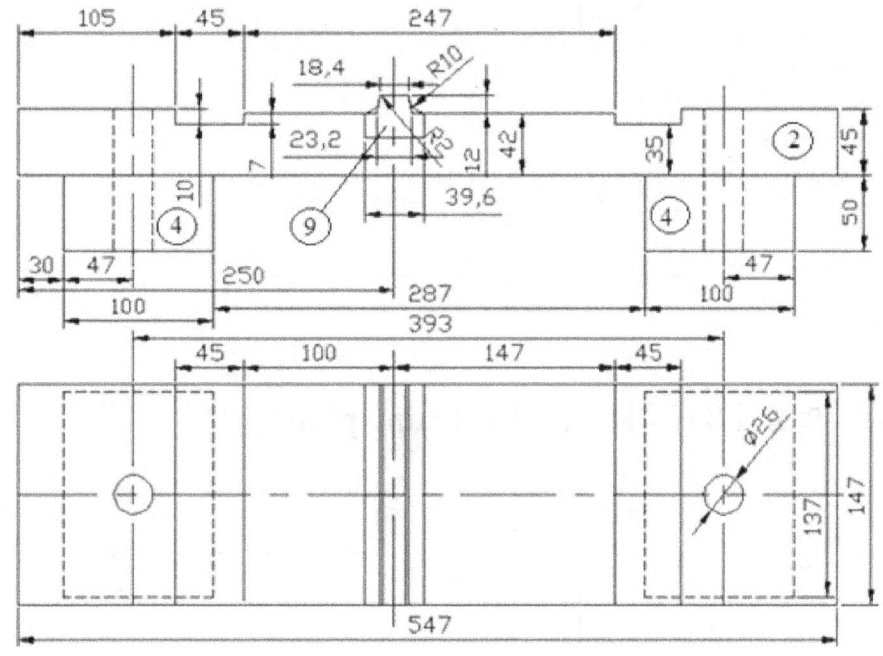

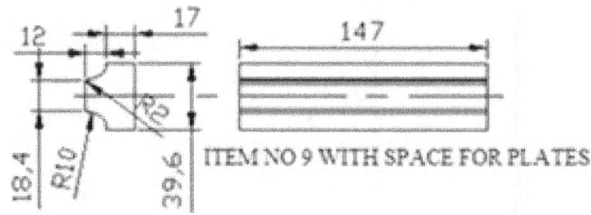

Assembly drawing of side wall (item No – 3), Battens (item No – 5) Loose pieces (item No – 6, 7 and 8):

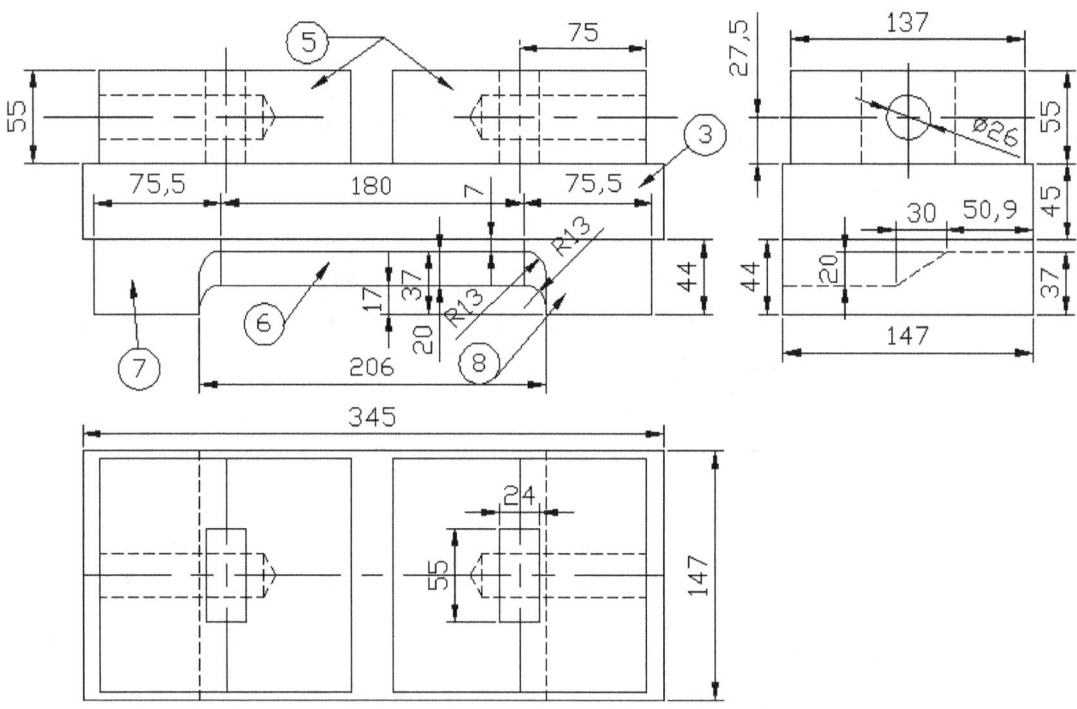

Item No – 6, 7 and 8 with space for fitting plate:

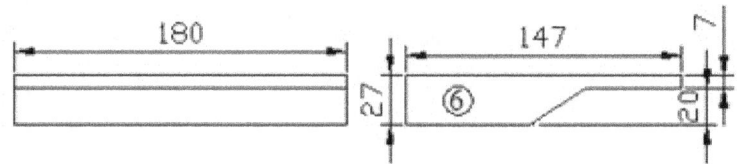

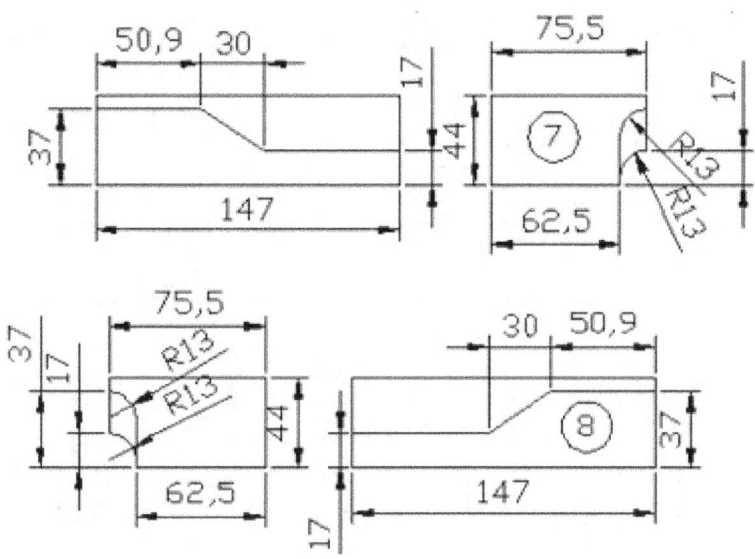

Side liner (Item – 3A), Loose Piece (Item – 10) and Side Battens (Item – 5) with space for fitting Plates:

In item – 10 the measurement 10 mm will become 7 after removal of 3 mm thick wood from surface

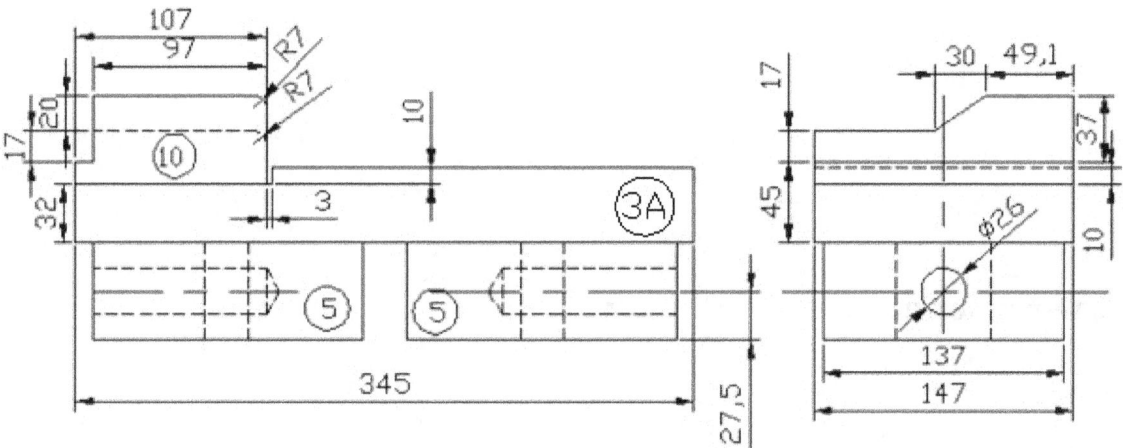

Item No – 3A, Item No – 10 and Item No – 5 with space for fitting plates:

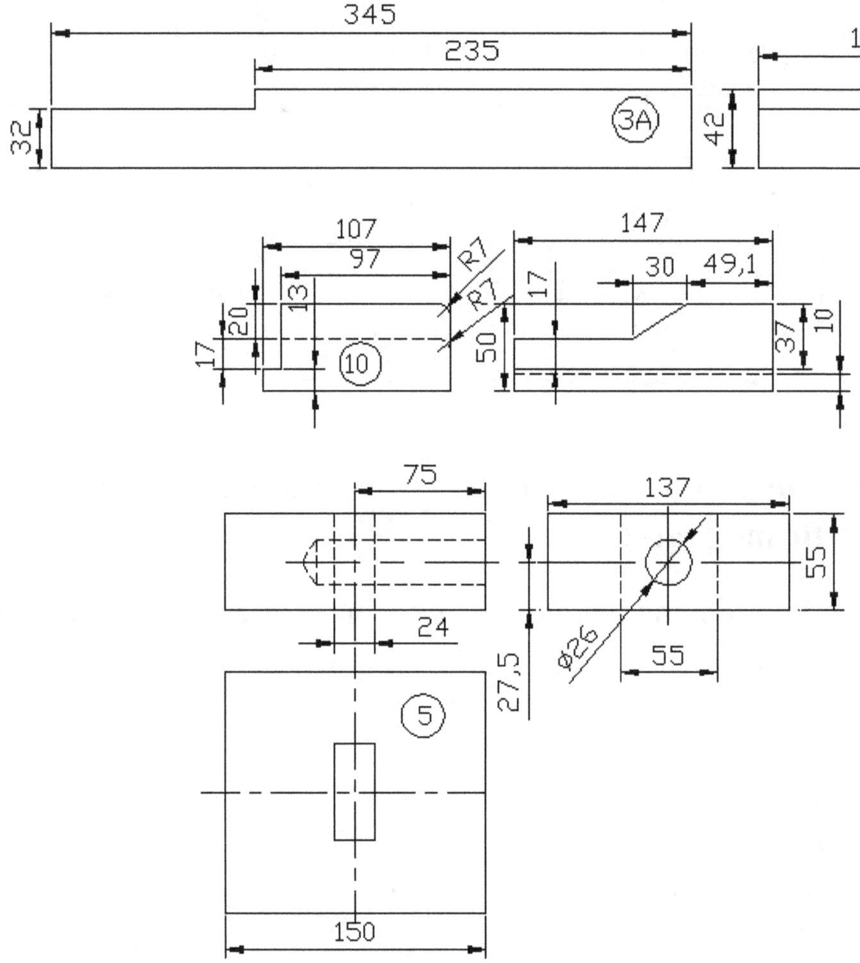

Fasteners for detachable assembly:

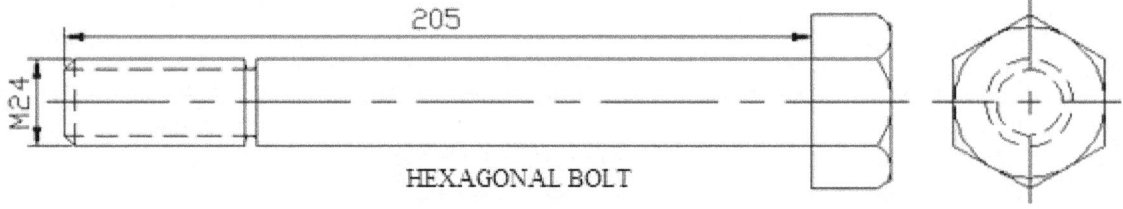

HEXAGONAL BOLT

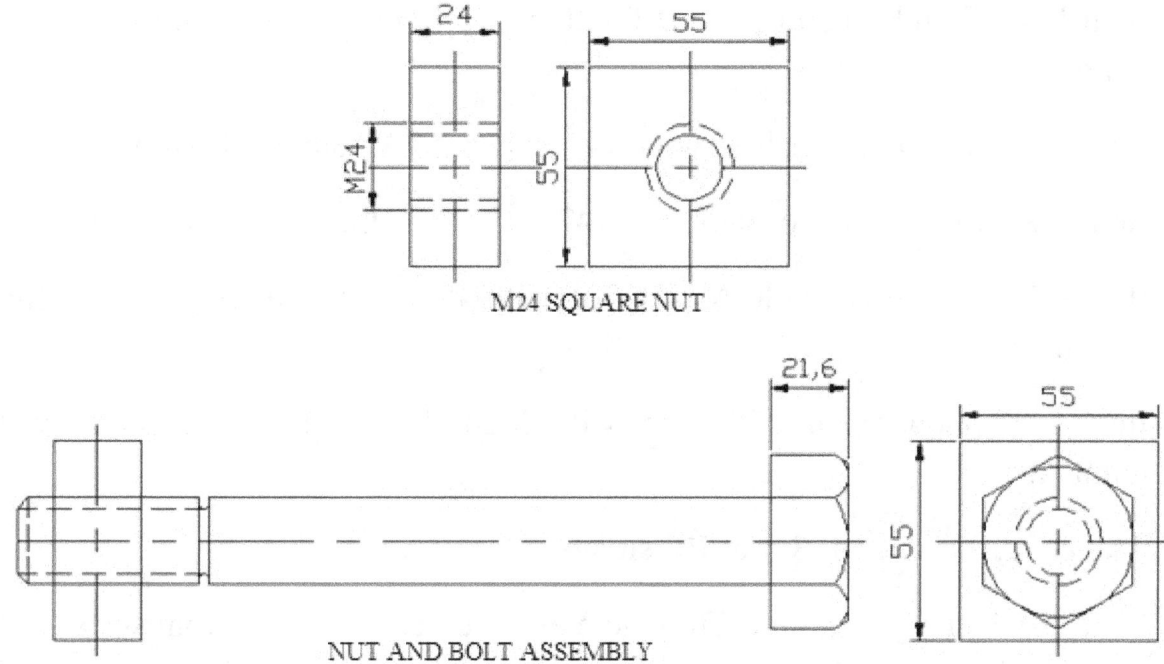

Step – 2 Material arrangements

Study the mould design with plate fitting and individual component drawings and find the size of wood and plates.

Item wise finished size of wood:

Item – 1 (End wall): 547 * 147 * 45, Quantity: 1 Number

Item – 2 (End wall): 547 * 147 * 45, quantity: 1 Number

Item – 3 (Side wall): 345 * 147 * 45, Quantity: 1 Number

Item – 3A (side wall): 345 * 147 * 45, Quantity: 1 Numbers

Item – 4 (End wall batten): 137 * 100 * 50, Quantity 4 Numbers

Item – 5 (Side wall batten): 137 * 150 * 55, Quantity 4 Numbers

Item – 6 (Loose piece for Item – 3): 180 * 147 * 27, Quantity 1 Number

Item No – 7 and 8 (Loose piece for Item – 3) 147 * 75.5 * 44, Quantity 2 Numbers

Item – 9 (loose piece for Item – 2): 147 * 39.6 * 29 Quantity 1 Number

Item – 10 (loose piece for Item 3A) 147 * 107 * 50 Quantity 1 Number

Item – 11 (Hexagonal bolt) M24 * 205 with 45 mm thread length - Quantity 4 numbers

Item – 12 (square nut) 55 * 55 with M24 thread in centre, Quantity 4 numbers.

The required sizes of 3 mm BP sheets:

Measure the length and breadth where ever you can measure from surface of individual items. Take the pieces of paper and place on inclined or curved surfaces and cut the papers. The papers must touch the surface perfectly to have length and breadth of plate for that surface.

Various sizes of screws for fitting with walls, battens and plates..

Step – 3 making Individual components

Process of making End Walls (Item No – 1 and 2)

Arrange wooden planks bigger than as mentioned in material list. Plane, make surfaces, sides and ends level, warpage free and right angle to each other. Use marking gauge, scriber and right angle to mark lines to cut for making grooves and space for plate fitting as in pictorial drawings. The markings on item number 1 and 2 must match with design lay out. The marking 1 and 2 on the front surface is the position of grove for side walls; marking 4 and 5 is place for 3 mm thick B.P. sheet. The position marked 3 on end wall item – 1 is the negative shape of tongue profile in the brick. The measurements have been transferred from layout design made with space for fitting plate. The radius profiles should be checked with gauge made for it. But Position marked 3 on end wall item – 2 is the groove for fitting loose piece item – 9. Use carpentry chisels and gauge chisel to cut the profiles.

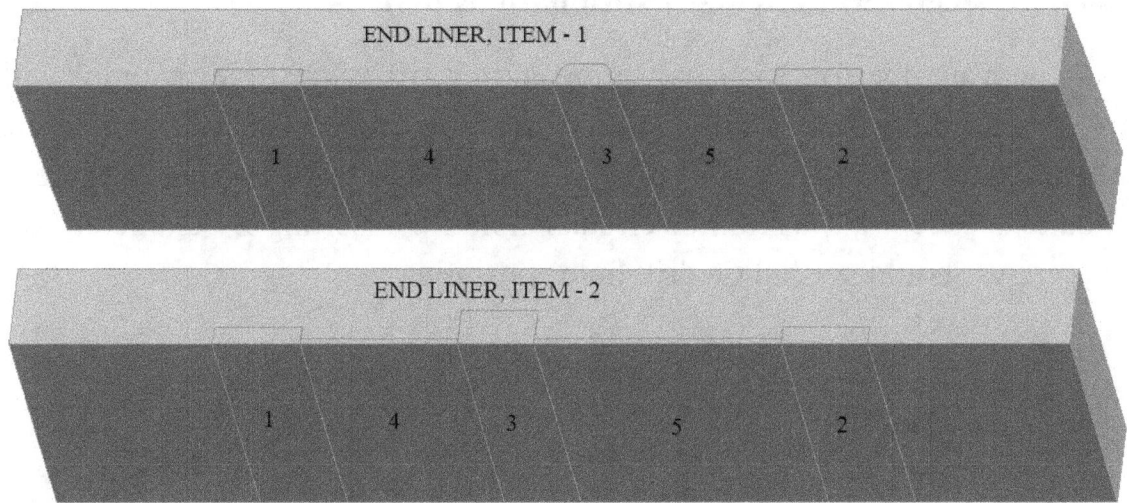

Process of making End Walls batten (Item No – 4)

Arrange wooden pieces of required size, Mark the required length, width and thickness; plane sides, surfaces and ends with planer and make right angle to each other.

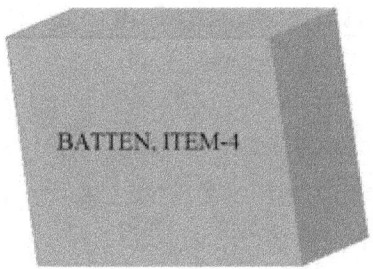

Process of making End Walls loose piece (Item No –9)

Arrange wood of required size, make surfaces and sides plane and transfer measurement from lay-out drawing. Cut the profile with suitable carpentry tools and check the profiles with gauge of required shape and size

Assembly of end walls item – 1 with battens item - 4:

Mark item number on each part after cutting off removable material with suitable carpentry tools. Apply wood adhesive on mating surfaces of battens and walls and fix with suitable size of screws. Mark the position of screws on battens except at position of circular hole for inserting 24 diameter bolt. The screws should not fall on drill holes position.

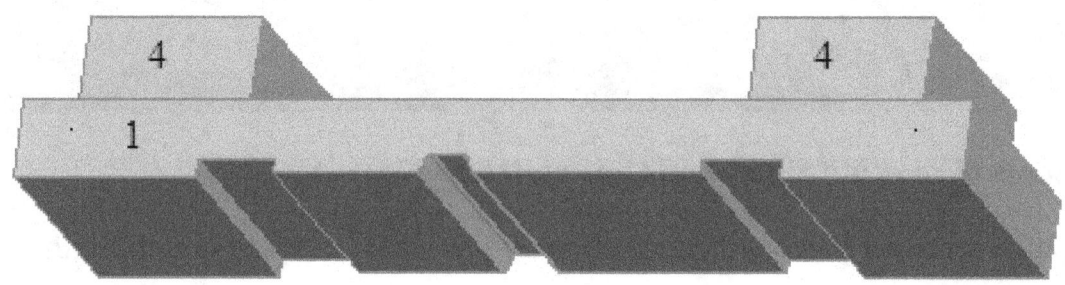

Marking for screws positions on surfaces of B.P. Sheet for fixing with item – 1:

The plate is made in 3 pieces and counter holes positions are marked as indicated in drawing. Make countersink drills matching with sizes of screws and fix on the surfaces at right positions.

Note: If plate fitting is difficult in groove, it is better to make loose piece with steel or aluminium and fix in position after making groove of matching size.

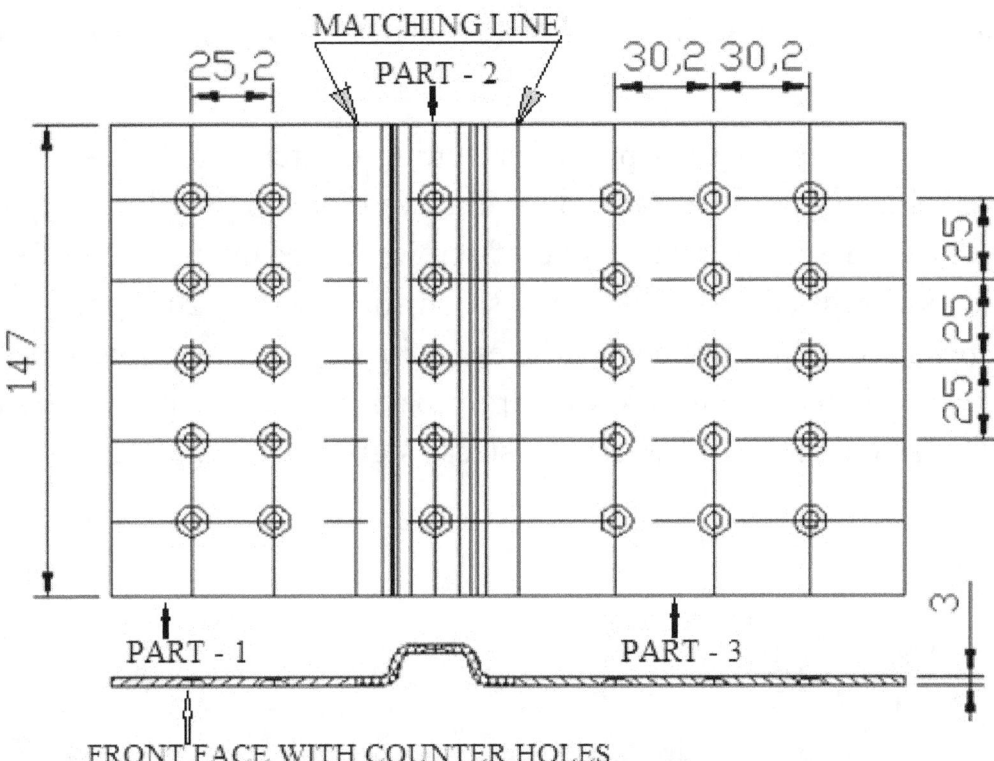

End wall item number – 1 with plates:

The screws have been fitted at 3 locations where countersink hole is not visible. Fix screws in all holes to attach with wood.

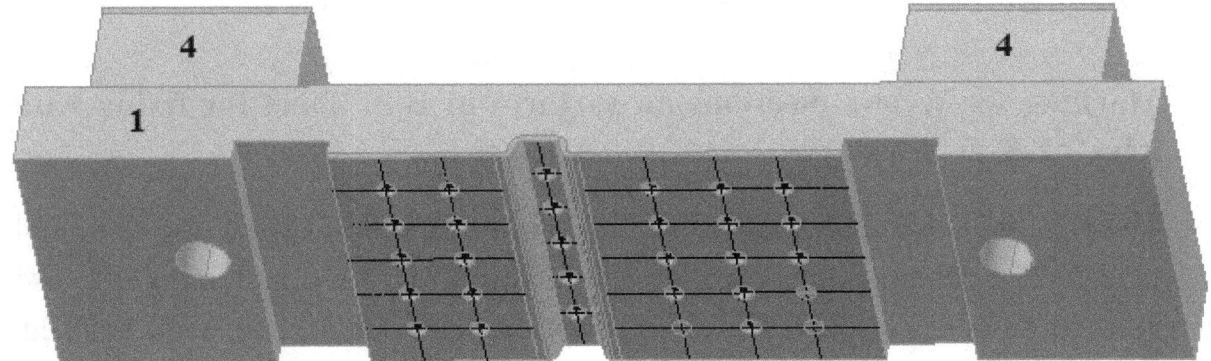

Assembly of end walls item – 2 with battens item – 4 and loose piece item -9:

Mark item number on each part after cutting off removable material with suitable carpentry tools. Apply wood adhesive on mating surfaces of battens and walls and fix with suitable size of screws. Mark the position of screws on battens except at circular holes position for inserting 24 diameter bolts. The screws should not fall on drill holes position. Insert loose piece item – 9 in groove of item – 2 and fix with from back face of item – 2. It will be better to make item 9 with steel with sizes as given in mould design without plate

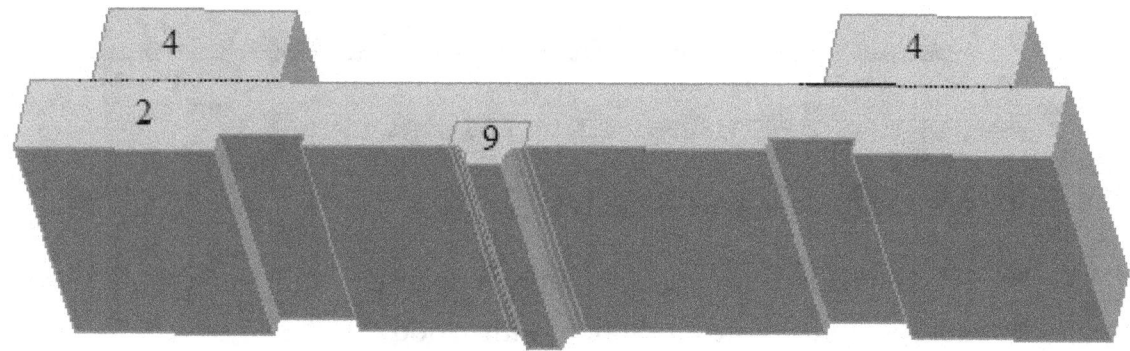

Marking for screws positions on surfaces of B.P. Sheet for fixing with item – 2 & 9:

The B.P. sheet for fixing on surface of item – 2 and 9 has been designed in 4 pieces. The counter holes position and drill holes with countersink are also marked in drawing. The matching lines of plates also have been marked. Prepare the sheets and fix on surface.

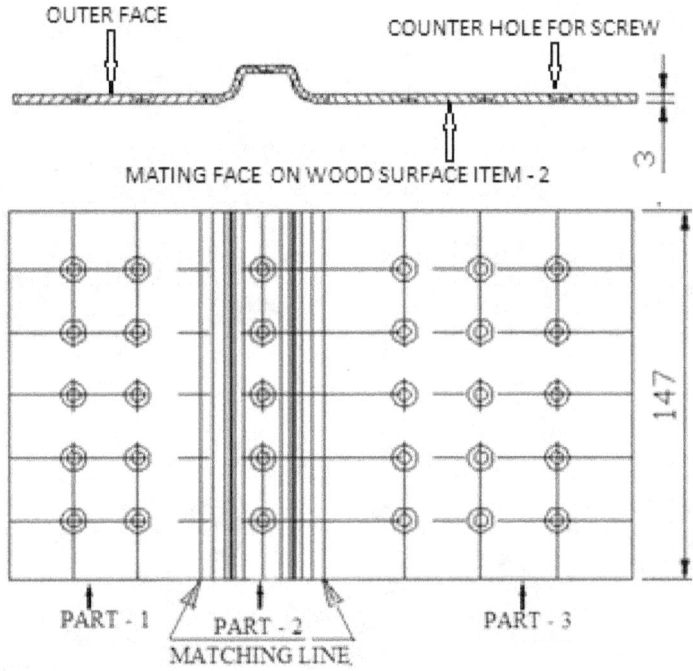

Assembly of item – 2, 4 and 9 with B.P. sheets fitted on working surface of mould:

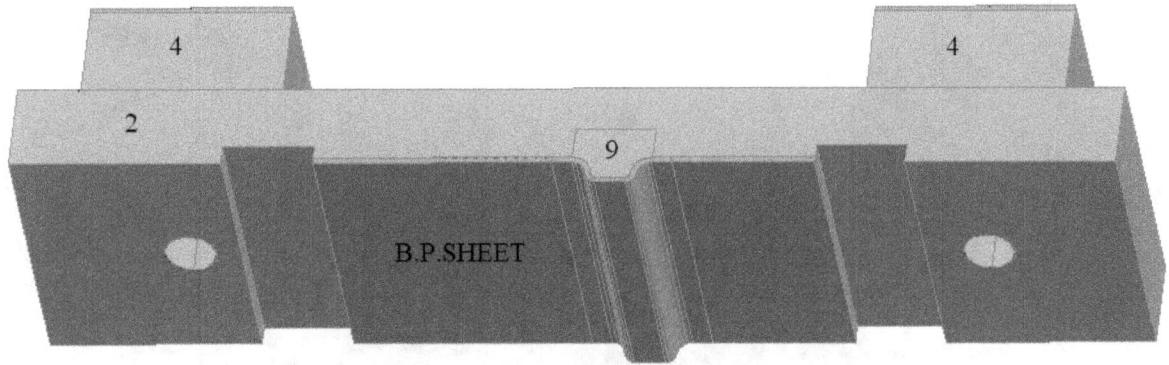

Marking to cut profile of loose pieces item number – 6, 7 & 8:

The item number – 7 & 8 are rotated for better understanding of marking on two end surfaces. The surface A and B are opposite end faces of the same loose piece. Transfer the measurements from layout to mark the removable area of wood. It has been done with suitable measuring and marking tools.

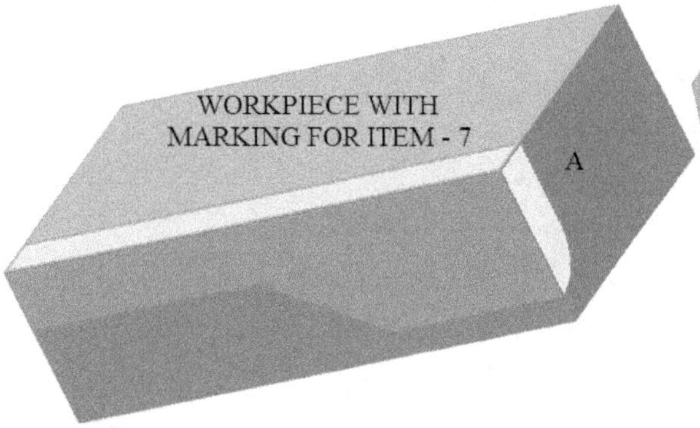
WORKPIECE WITH
MARKING FOR ITEM - 7
A

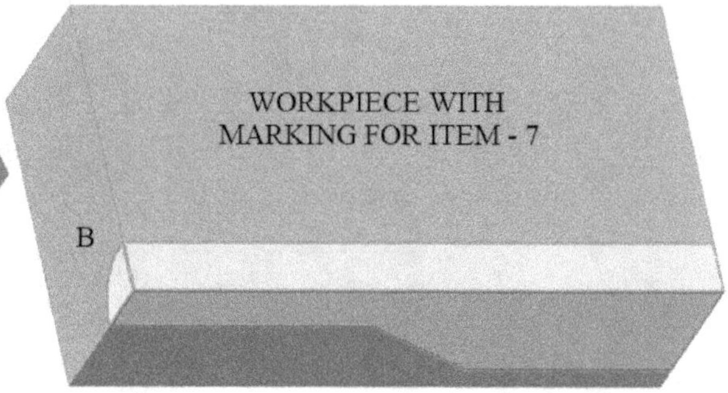
WORKPIECE WITH
MARKING FOR ITEM - 7
B

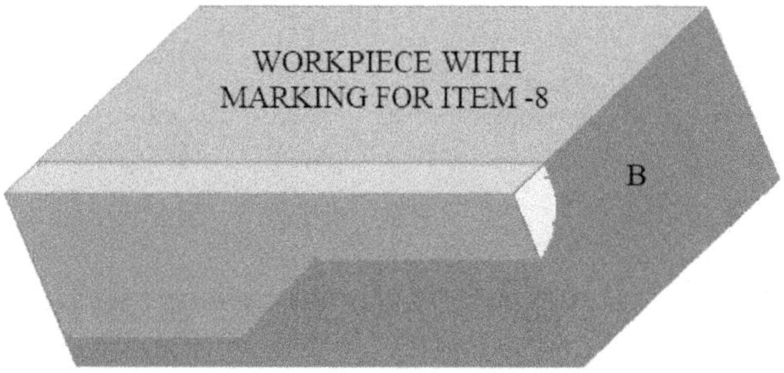
WORKPIECE WITH
MARKING FOR ITEM -8
B

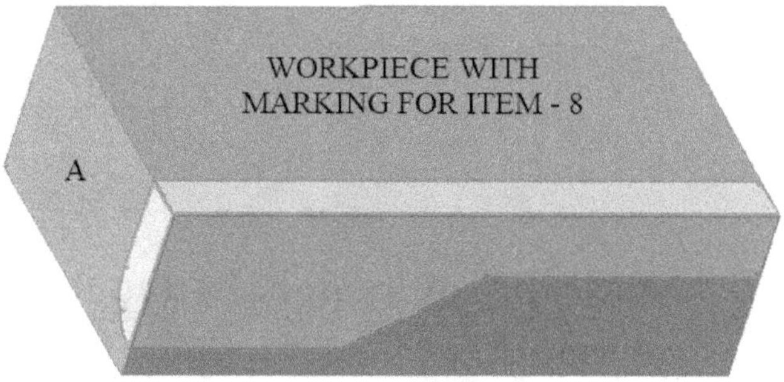
WORKPIECE WITH
MARKING FOR ITEM - 8
A

WORKPIECE WITH
MARKING FOR ITEM - 6

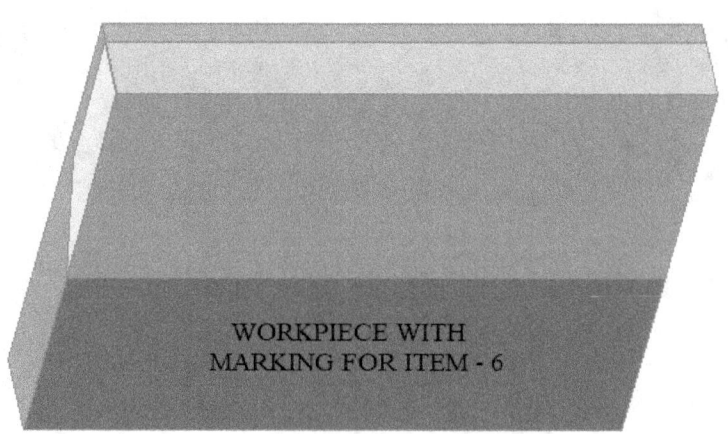

Shape of item – 6, 7 & 8 after cutting off marked area:

Cut off the wood with appropriate carpentry tools from marked area as indicated in above pictorial views. Place 3 loose pieces on level surface and see the assembly for perfection. It should look as in pictorial view of assembly drawn here.

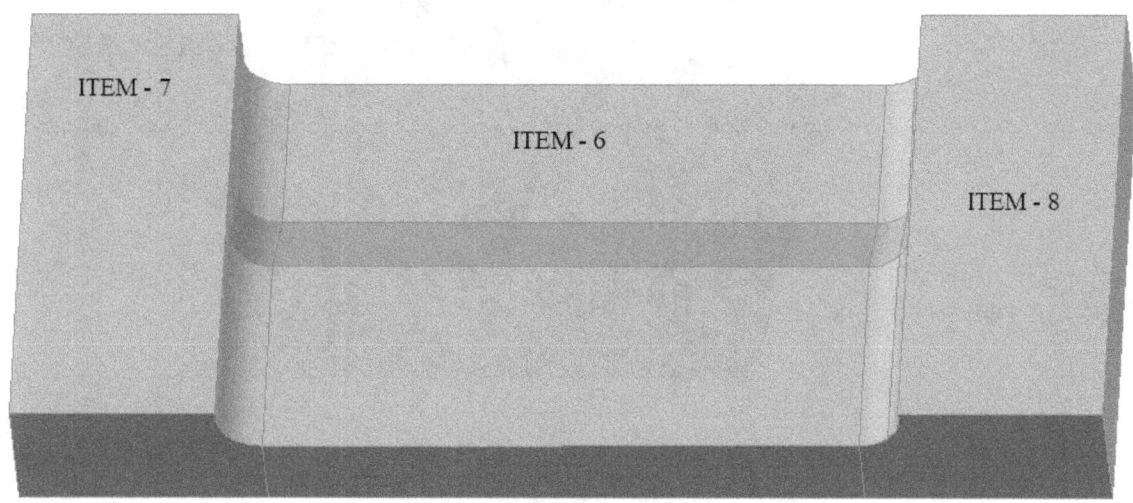

Manufacturing Side wall battens:

You need four pieces of side wall battens of size and shape with some special features. Arrange wood of required size, plane, make surfaces smooth, mark and cut to size. Mark the position and size of pocket for inserting square nut on front and back surfaces. Mark centre lines on one end face and a circle of 26 Ø at intersection point. Cut the pocket on surface of size 55 * 24 * 55 deep. Make a drill hole of 110 mm length at intersection of centre lines on left side face. You have to insert 205 mm long bolts in the assembly to fix with square nuts. The length of the hole in the side wall batten should be 45 + 50 + 3 + length of hole in the side wall batten= 205. Length of hole in the side wall batten = 205 – 98 = 107. The length of hole must be more than 107mm. These measurements relates to assembly drawing of mould

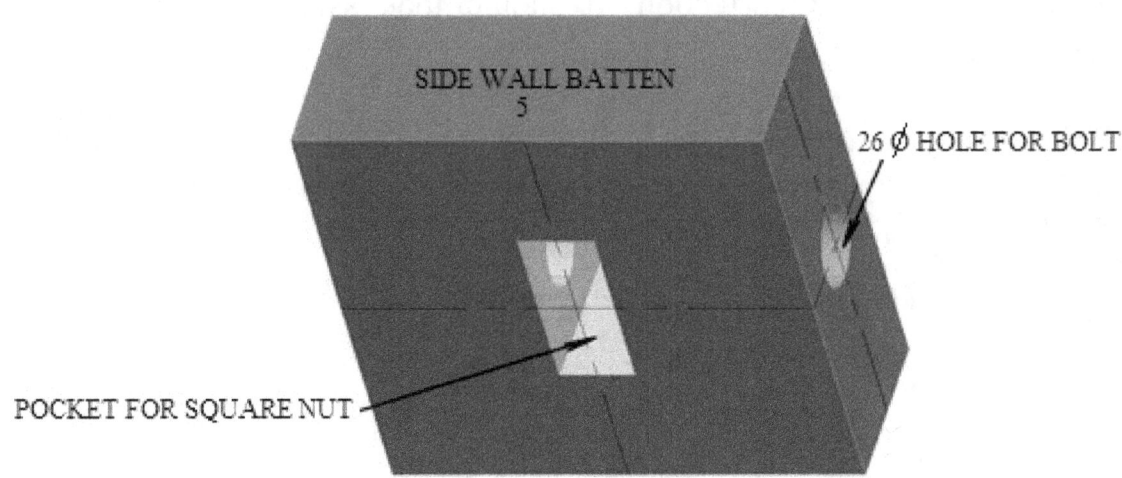

Assembly of side wall item – 3, battens and loose pieces:

Assemble the side wall battens on the back face of side walls. Mark the position of screws on the surface of battens, make slightly smaller drill hole than the shank dimeter of screws and fix on the back face of wall. Apply wood glue (adhesive) in between the mating surfaces of wood. There should not be screws in the location of 26 Ø hole. The size of screws should be large enough to hold wall strongly.

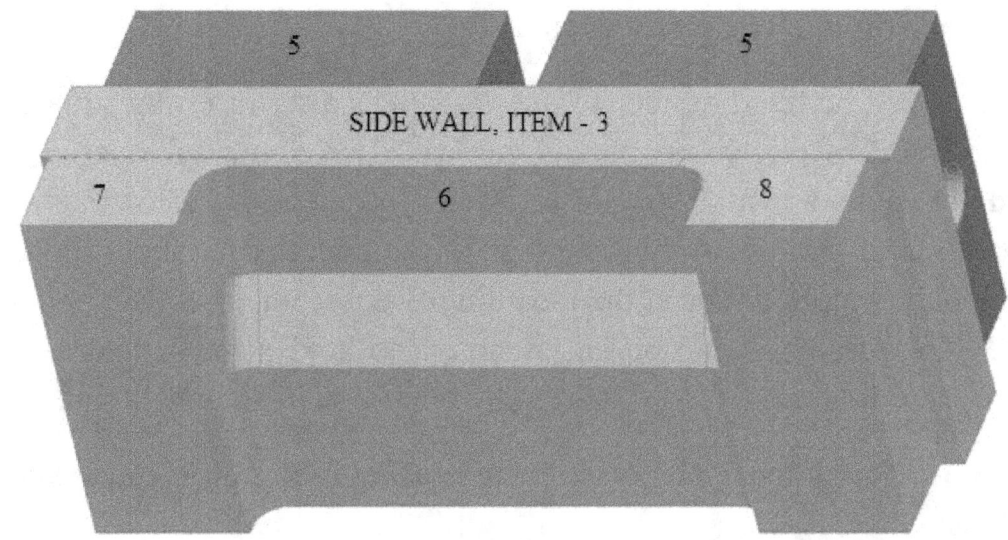

The back face of battens brought in front with wall and loose piece to focus on slot for nut and holes for bolt.

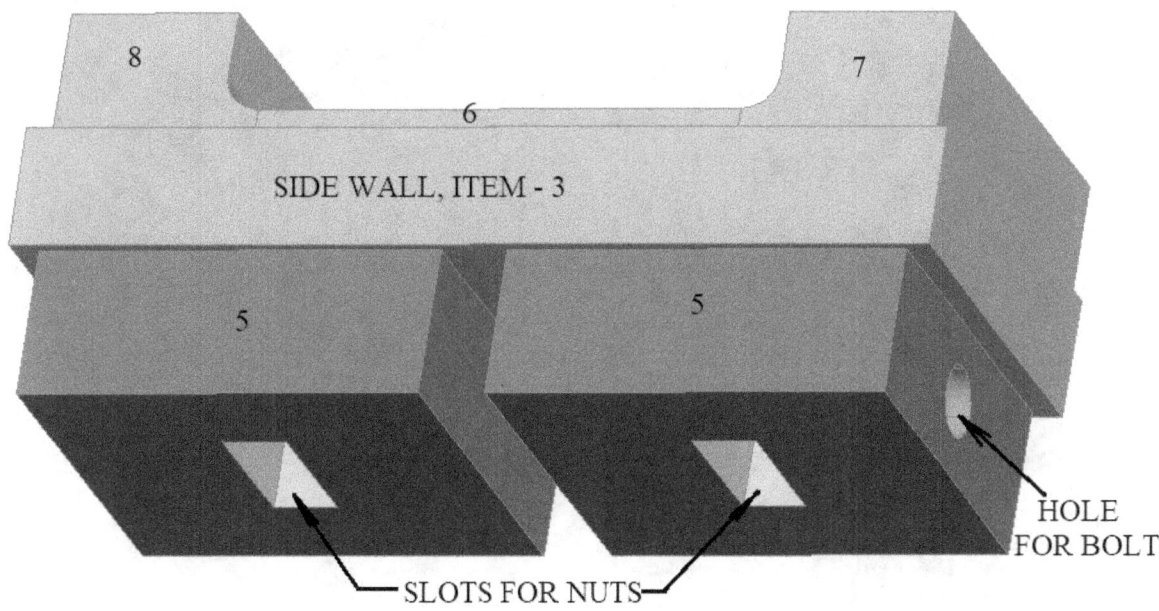

Plate fitting on back face of battens:

The plates will be fitted on the surface of battens after inserting square nuts. Mark location of screws on plate's surface. The centre of holes on plate should not coincide with centre of screws on battens (screws on battens and screws inserted from plate should not be at same location). The screws fixed through plate should not be at the location of square nut and 26 Ø hole in the batten. The screws should be large enough to hold plate battens and wall strongly. Screws thread should hold more than 70% of side wall thickness.

Side wall – 3 with loose pieces & battens after fixing plates in respective positions:

Divide plates into convenient numbers considering ease of bending. Bend, match edge to edge, corner to corner at radial potion, drill countersink holes and fix on the surface.

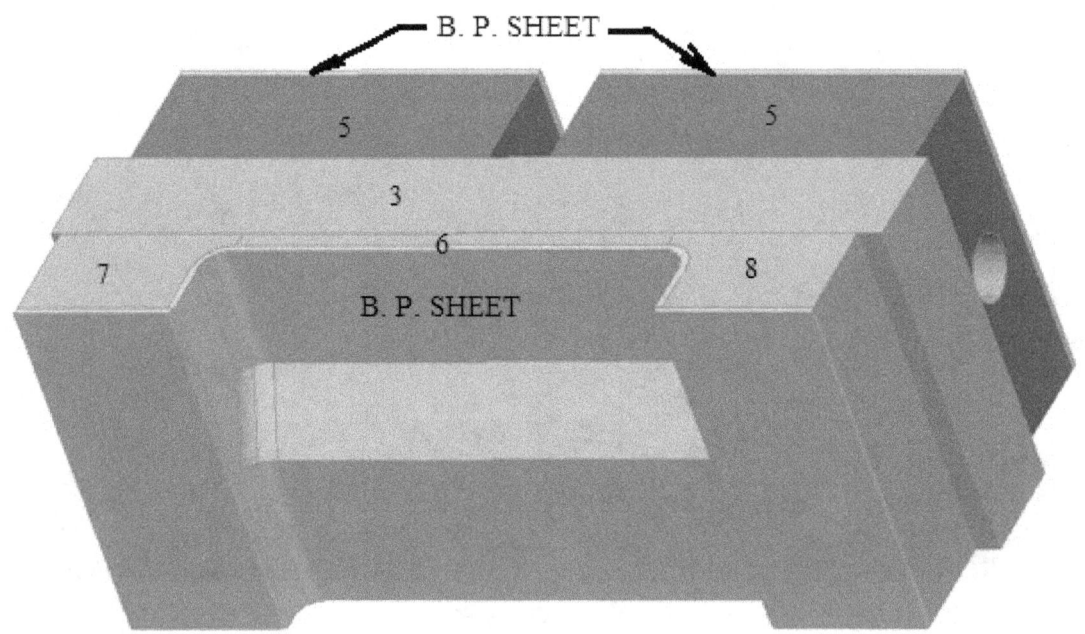

Side wall item – 3 A and loose piece item – 10:

Arrange wood of required size as in material list make it ready with the help of marking and measuring tools as earlier. Remove the wood with cutting tools according to marking and get the shape as in pictorial views. The loose piece has been placed in position on side wall without plate fitting.

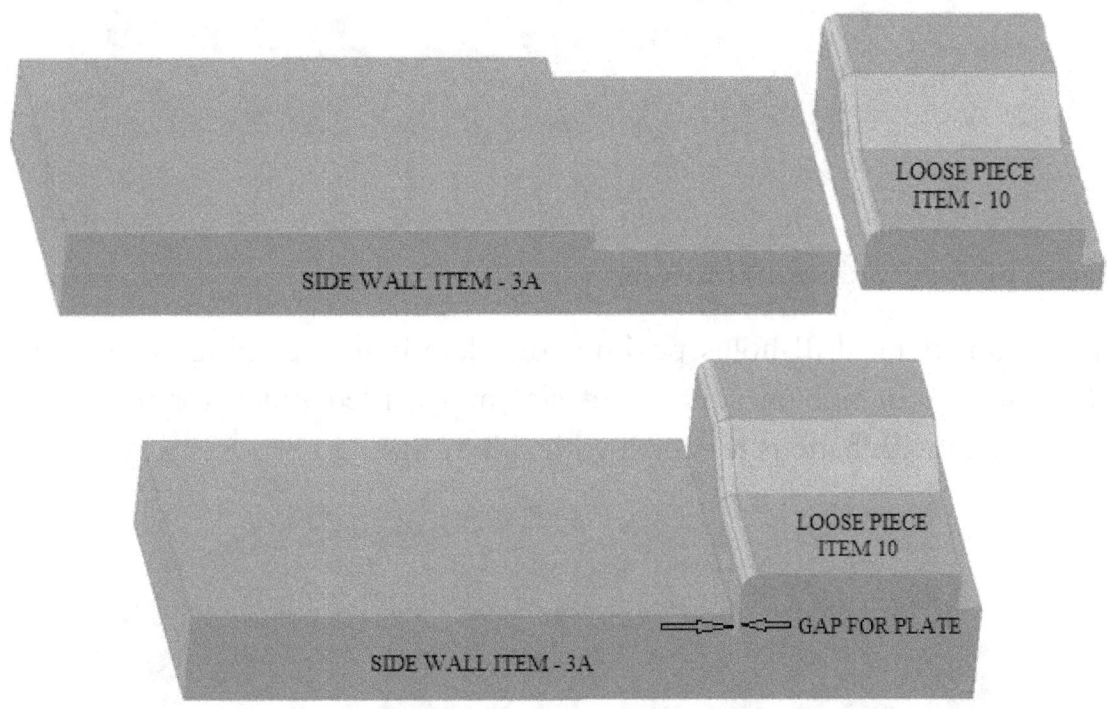

Plate fitting on surface of item – 3 A and loose piece item – 10:

Cut the plate of require size as per convenient and facilities available for bending at radial portion. You can make in 3 pieces one for inclined surface and two for straight plain surfaces. Mark countersink hole centres on outer surface of plates (opposite to contact surfaces of wood). Make countersink drill holes on drill machines or by portable drill. While drilling countersink holes provide wooden support under the plate to prevent possibilities of bending. The plates should seat perfectly on surface of wood after fitting screws. The process of marking and drilling countersink holes to fit screws has been explained earlier.

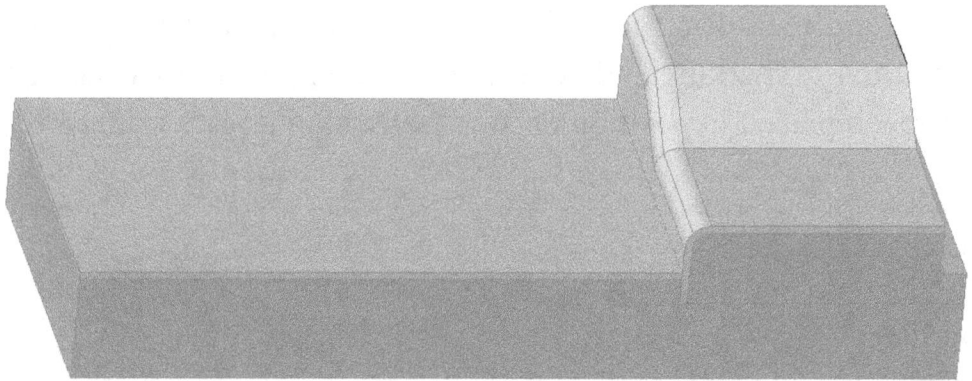

Fitting plate on side wall battens:

Mark countersink drill holes positons on plate before drilling. The counter holes should not fall on square nut slot and drilled holes for nut and on screws fitted with battens to attach with side wall.

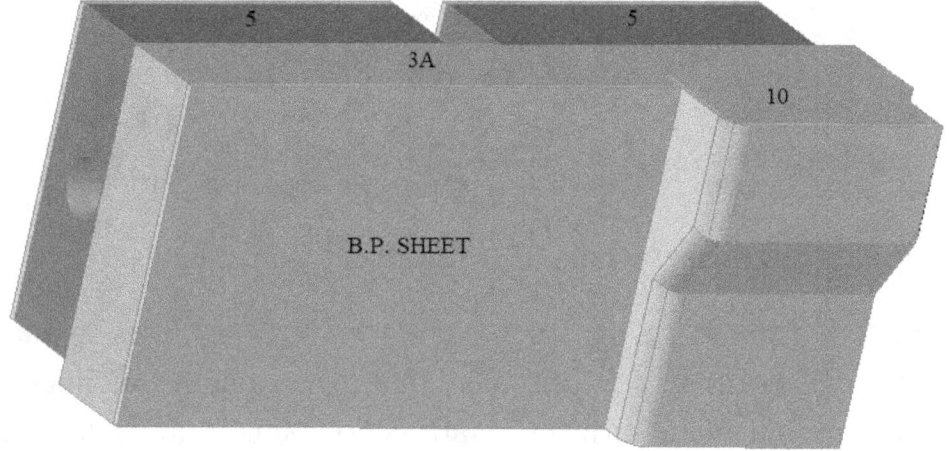

Assembly of side wall and end walls with bolts:

Assemble the walls and fix the bolts in square nuts through end wall battens.

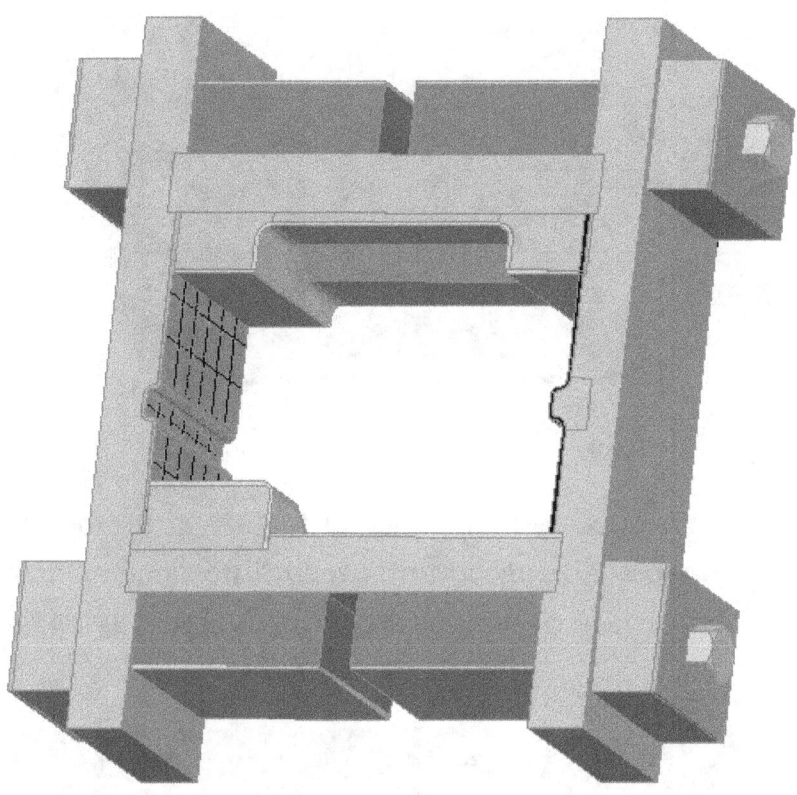

Note: The tongue and groove profile can be made with steel if there is difficulty in bending plates with smaller size radius.

Manufacturing of bottom board:

The length and width of bottom board should be suitable to accommodate walls and battens of mould. The board is composed of top surface and battens to hold the top surface planks with screws. Arrange wooden planks of required size for top surface and battens for joining and supporting the planks. Plane and make warpage free level surfaces of planks and battens. Join them with screws. You can fix plate on surface of board if it is for moulding large quantity. The bottom die is fitted on top surface of bottom board.

Manufacturing of bottom die:

The bottom die is required to make special feature at the bottom surface of the brick. The peripheral size of bottom die will be same as outer peripheral shape of brick. The surface of bottom die will have negative shape of tongue in the brick.

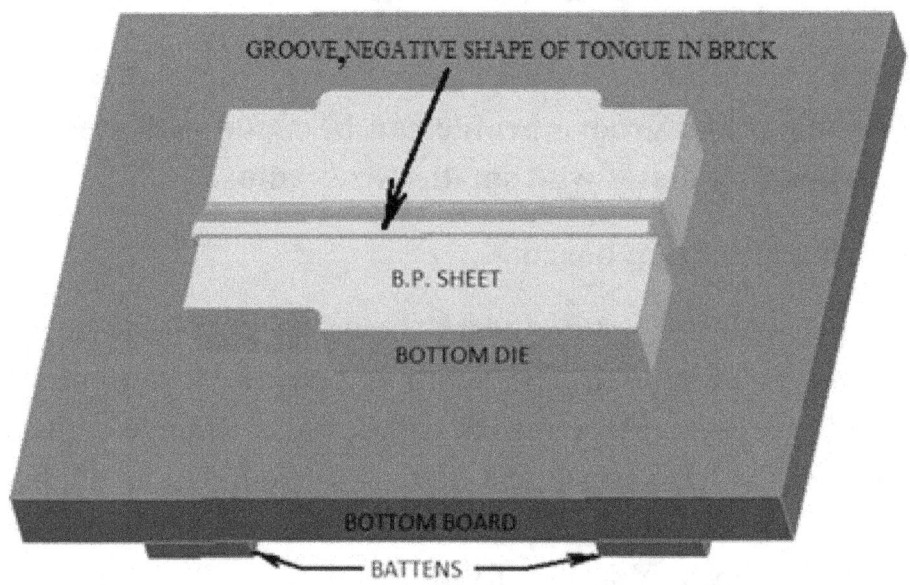

Formation of Groove on top surface of brick:

The shape and size of groove on top surface of brick will be made with top die gauge. This top die gauge will form groove when set on the surface of the mould with top cover. Measure the length along groove cut on to surface of the mould. The length should be more than the length from one end wall to another opposite end wall. The width of the plank will be more than tongue profile (negative shape of brick groove). Arrange the wood of suitable thickness, length and width and cut the shape and profile of tongue. Plate on surface also is to be fitted considering number of pieces to be moulded.

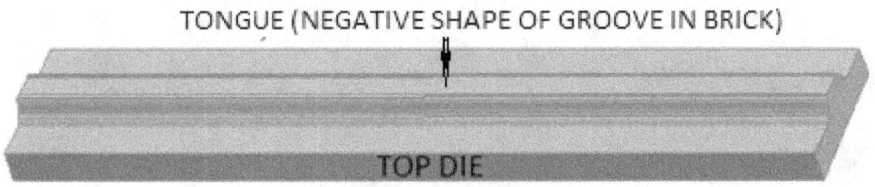

Top Cover Plate (Frame): The top cover frame is made of hard wood. It is used to protect the top surface of mould from damage by pneumatic rammer while compressing mixture in the mould cavity. The projected portion of top die gauge will fit in the groove of top frame.

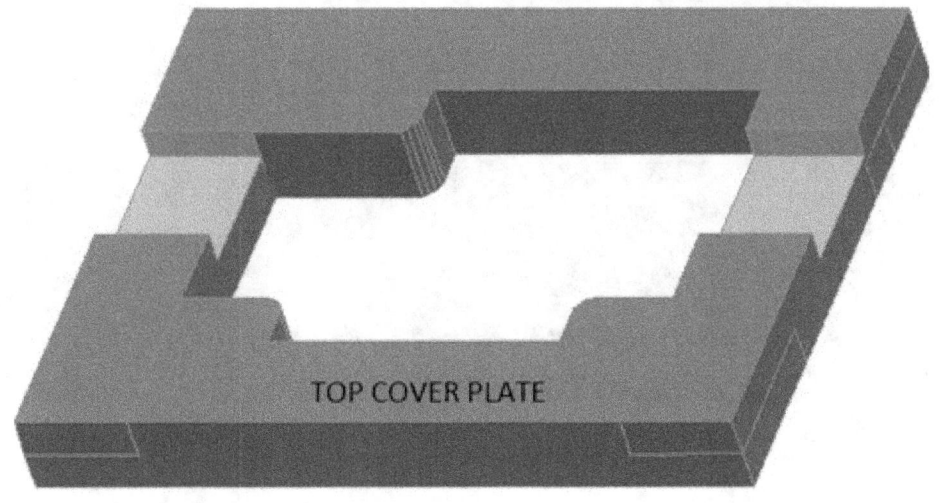

Cutting Groove profile of brick on top surface of mould:

Mark the shape and size of groove profile in correct location on top surface of mould by transferring measurement from design layout. Remove material and finish the shape with appropriate tools.

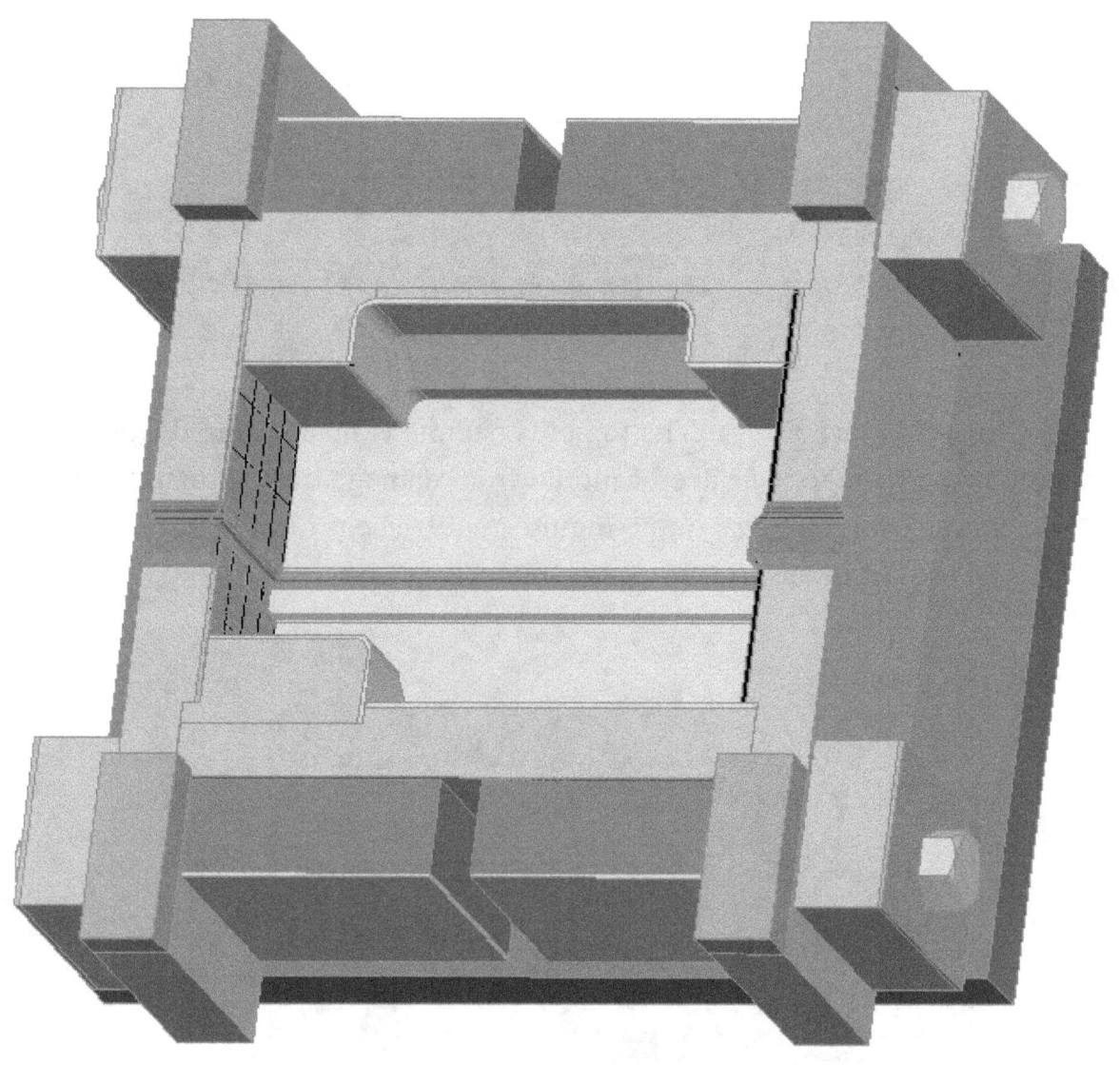

Note: The battens of the board are not visible due to rotation and inclination angle of pictures in the under given drawings. See the mould assembly with top die and top frame ready for moulding.

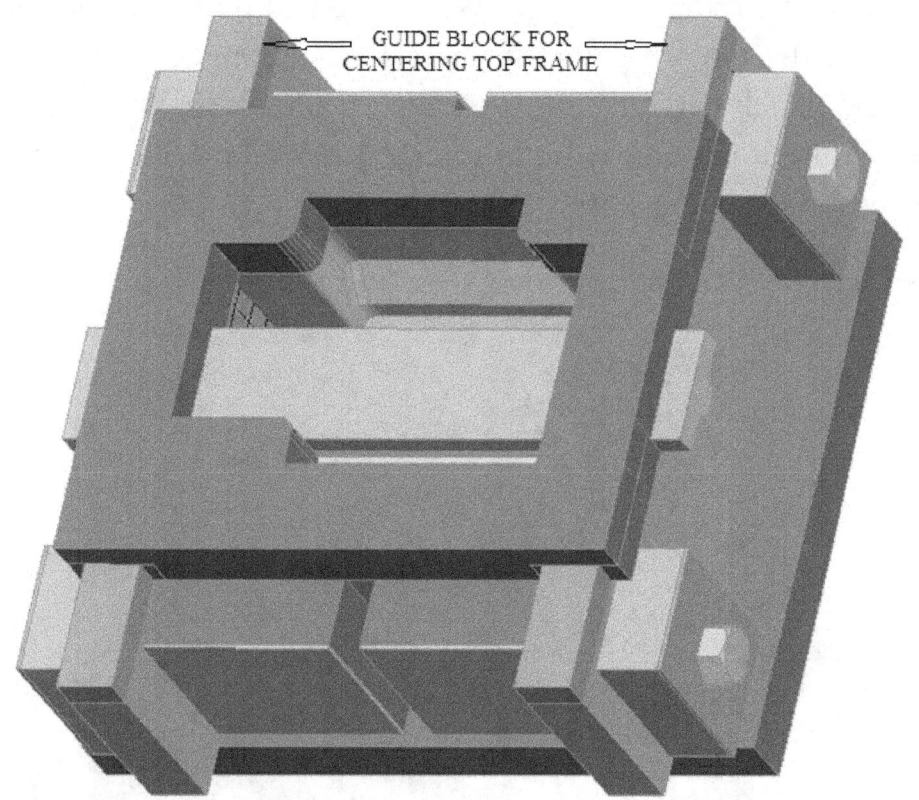

The process of making brick:

Place the mould assembly on brick moulding table. Place top cover without top die gauge on the upper surface of mould assembly. Align the mould cavity with top cover cavity. All the plates inside the cavity must be under the top cover to protect from the impact of rammer while pressing the mixture. The movement of pneumatic rammer is under manual control, there may be chance of damage if plates are not covered by top cover. Clamp the top cover in position as illustrated in drawing or with available mechanism in moulding shop. Prepare the brick mixture to have required bulk density and quality to suit the customer's need. Weigh the mixture on balance to get required bulk density. Compress the mixture in the cavity of mould with pneumatic rammer. Pack the mixture uniformly in the entire cavity under uniform pressure to have uniform density in the brick. First ram the mixture in the cavity without top die gauge. When rammed mass will be near to top surface, insert the top die gauge in the slot. The tongue profile of top die gauge should match perfectly in the groove made on top surface of mould. Ram the mixture in the cavity up to top surface of mould and extra mixture in the top cover above the surface of the mould.

Mould with top cover and top die gauge on moulding table:

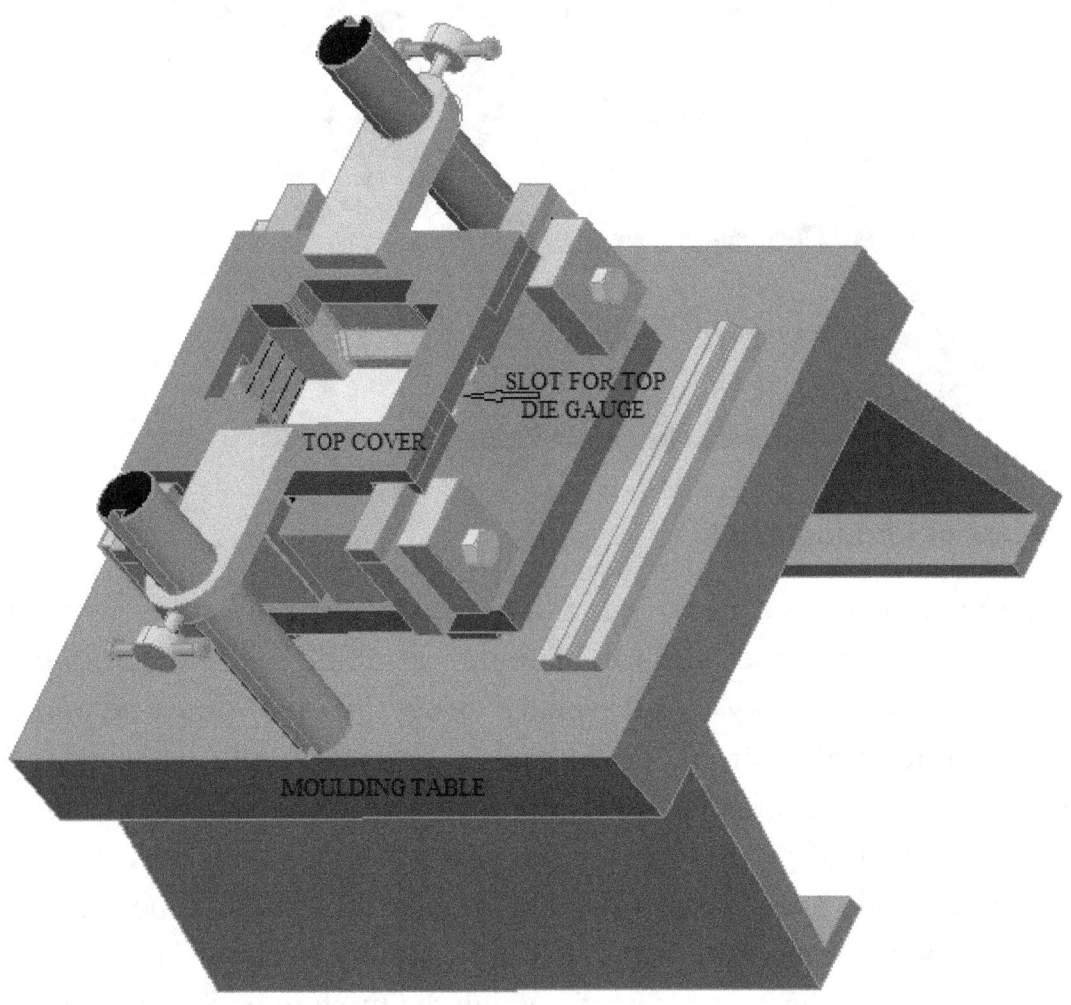

Process of taking out from moulding table: First remove the clamp and take out mould with brick putting a crane rope under the bottom board and in between the board battens. Place it on level floor. Take off top cover with top die carefully from mould. Cut off extra mixture rammed on surface and polish it. See the preceding pictorial drawings.

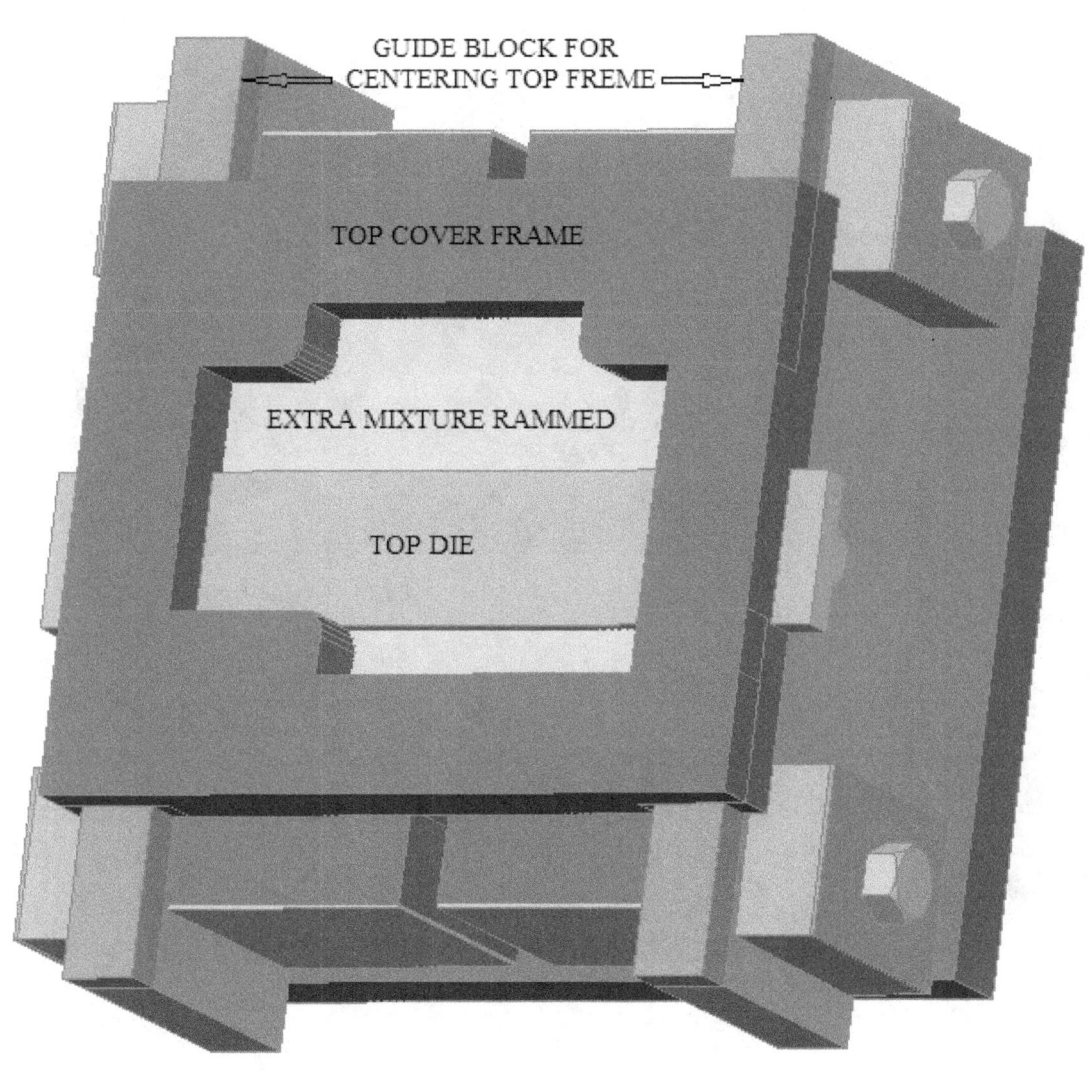

Process of removing extra rammed mixture:

Take suitable size of knife edge steel bar; scrap the rammed mass slowly by pushing with hand pressure. With rubbing and push action forward the extra mixture will come out from the surface. Check the surface with straight edge and polish it.

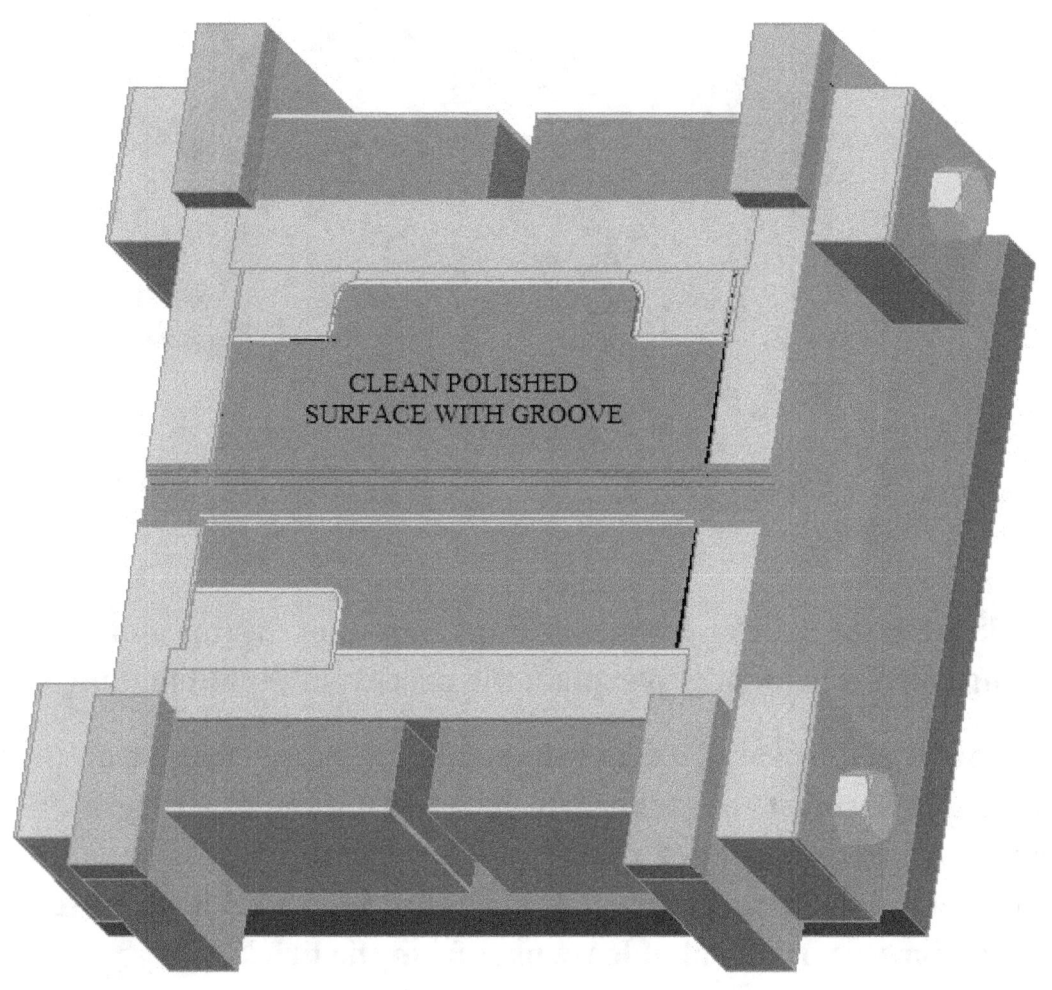

Dismantling process of mould to take out brick:

Hit on two ends of one end wall surface gently with wooden mallet after removing all hexagonal bolts and remove the end wall. Similarly remove opposite side's end wall. Pull away two side walls also from the surface of brick. Place the brick on pallet and send to drawer.

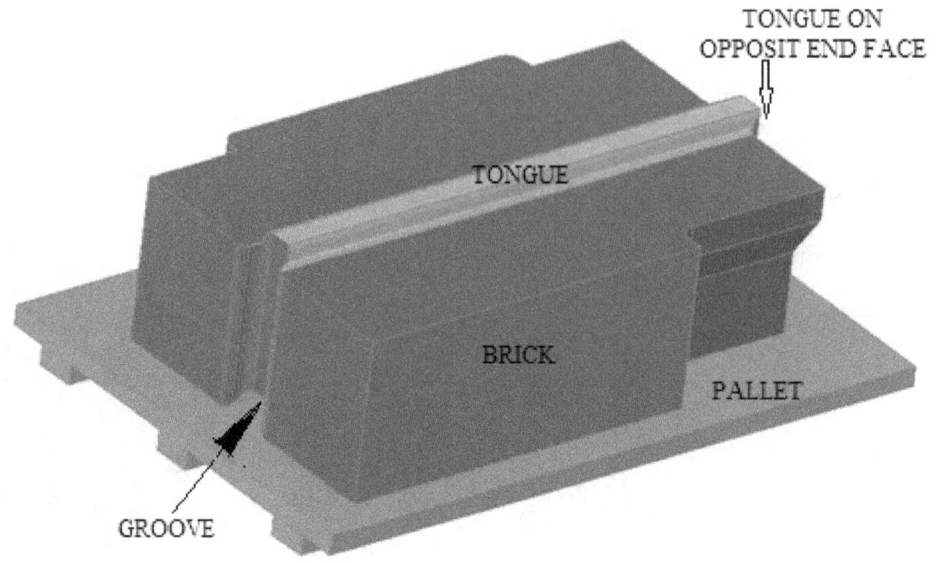

Note:

The sketches of bricks in the example are not usable. The shape drawings have been made to explain the process of mould making.

The process will be same for any shape having simple feature on the top surface. Care must be taken to have sustainable strength in the assembly against impact of rammer. The assembly and dismantling of mould components must be easy. If required suitable draft can be provided for easy removal of loose part from the brick.

The mould design is based on brick lay-out with allowance. The component (individual item) size must be developed from mould assembly drawing.

Chapter – 8

Questions for Skill test

The answer to all questions listed here are available in chapter – 9

Questions on mould making

1. What is mould?
2. What is necessity of wooden mould?
3. What are most important required skills of a mould maker?
4. What are the precautionary measures to be taken in mould making?
5. What are basic differences between mould making and carpentry work?
6. Is there any chance of side button coming out and dismantle the mould during ramming operation of mixture?

Question on brick moulding:

1. What is moulding?
2. What is reciprocating motion?
3. Why top cover frame is required?
4. Why extra brick mixture is to be rammed above the mould top surface?
5. Do the extra mixture rammed above the top surface of mould is included in total weight of mixture to be rammed?
6. What is bulk density?
7. How to calculate weight of brick mixture considering bulk density?

Chapter – 9

Answers to questions in chapter 8

Answers on mould making

1. Mould is an object with outer and inner shape. The inner shape is the negative shape of the brick that is to be made. The outer shape is made by mould maker according to design of detachable assembly. In complex shape having feature on all surfaces of brick, the upper and lower features are provided with upper and lower dies in addition to peripheral shape in inner cavity.

2. The wooden moulds are necessary to produce complicated shape bricks that cannot be made on press with steel moulds.

   Factors involved in consideration for making wooden mould
   - The bricks that can't be ejected from fixed mould after pressing are considered for making wooden moulds
   - Large size bricks that can't be made on press are considered
   - When few bricks are to be produced for customer

3. Important skill required by mould maker
   - Clear concept of engineering drawing. He should be able to read and draw drawings as he has to make layout and develop individual assembly parts of mould on layout board.
   - He should have knowledge of bench work such as drilling tapping, filling and plate fitting. He should also have sufficient knowledge of measuring, marking and cutting tools including precision measuring. If interested to learn precision measuring read my book "Jigs, Fixtures and precision measuring Instruments".

- He should be able to weld and do gas cutting if required.

4 Precautionary measures
- He must check layout with brick drawing for any mistakes
- Positioning of loose pieces should be in correct place. It must be checked before plate fitting work starts.
- Screw position on battens to attach with walls must be marked with specific measurements. The screw must be long and strong enough to hold wall with batten strongly.
- The position of screws on plate that is to be fitted on batten should be different than the screws fitted on battens to attach wall. These screws must hold plate batten and wall together.
- Necessary draft is to be provided where ever essential.
- The brick tolerance provided by customer on bricks must be maintained.

5 Basic difference in mould making and carpentry work:

The mould making is bit precise work than carpentry work. The mould making requires deep knowledge of engineering drawing whereas carpentry work needs only primary knowledge. The important considerations in both fields are as under:

Important consideration in mould making
- Tolerance on brick dimensions,
- Perfection in shape of brick,
- Strength of mould,
- Easy assembly and dismantling to remove brick after moulding,
- Perfection in assembly and strength of tie bolts to maintain rigidity in mould during ramming the brick mixture

Important considerations in carpentry work.
- Outlook and finish

- Strength
- Customer's need or utility

6   The mould may dismantle during ramming operation if wall and battens ae not fastened strongly with screws of longer length.

Answer on brick moulding

1. Moulding is the process of making brick by compressing brick mixture inside the mould cavity with mechanical equipment such as, pneumatic force, hydraulic force, stamping force or vibratory force.
2. The reciprocating motion is a repetitive movement of movable part of a mechanical assembly. It is up – and – down or back – and – forward at regular interval of time in fixed cyclic order. The double acting stationary steam engine is an example that coverts reciprocating motion to rotary motion. The two opposite motion that comprise a single reciprocation cycle are called strokes.
3. The top cover frame is required to protect the top surface of mould from damage while ramming brick mixture with pneumatic rammer.
4. The extra brick mixture is rammed above top surface of mould to have uniform bulk density on upper layer of brick and polish smooth surface after cutting of extra mixture from top surface.
5. The total weight of mixture calculated considering volume and bulk density is packed in the cavity layer wise at regular interval. The mixture rammed above the top surface of mould in the top cover frame has extra weight that is not included in the total weight of brick.
6. The bulk density is the measure of the weight of a given volume of refractory brick. The refractory brick with higher bulk density is better in quality. The structure of brick with higher density is denser, resulting in better resistance to chemical attack, decreased metal penetration and better abrasion resistance.

7. $P = m/v$,

   p = density, m = mass of the brick and v = volume of brick

   Mass (weight) = $p*v = m$
   Volume = length * width * thickness
   When Bulk density of the brick is known you can calculate the weight of brick.

I would love to hear feedback and inputs from readers and practitioners. You can provide your inputs and feedback by emailing at reesaa@indiavivid.com or visiting www.reesaa.com

- Sheojee Prasad

www.ingramcontent.com/pod-product-compliance
Lightning Source LLC
Chambersburg PA
CBHW081002170526
45158CB00010B/2882